WIRELESS MESSAGING DEMYSTIFIED

McGRAW-HILL NETWORKING AND TELECOMMUNICATIONS

3G Wireless Demystified
802.11 Demystified
Bluetooth Demystified
CEBus Demystified
Computer Telephony Demystified
DVD Demystified
GPRS Demystified
MPEG-4 Demystified
SANs Demystified
SIP Demystified
SONET/SDH Demystified
Streaming Media Demystified
Video Compression Demystified
Videoconferencing Demystified
Wireless Data Demystified
Wireless Messaging Demystified
Wireless LANs Demystified

Wireless Messaging Demystified

Donald Longueuil

McGraw-Hill
New York Chicago San Francisco Lisbon
London Madrid Mexico City Milan New Delhi
San Juan Seoul Singapore Sydney Toronto

The McGraw-Hill Companies

Cataloging-in-Publication Data is on file with the Library of Congress.

1 2 3 4 5 6 7 8 9 0 DOC/DOC 0 9 8 7 6 5 4 3 2

ISBN 0-07-138629-7

The sponsoring editor for this book was Judy Bass and the production supervisor was Sherri Souffrance. It was set in New Century Schoolbook by MacAllister Publishing Services, LLC.

Printed and bound by RR Donnelley.

McGraw-Hill books are available at special quantity discounts to use as premiums and sales promotions, or for use in corporate training programs. For more information, please write to the Director of Special Sales, Professional Publishing, McGraw-Hill, Two Penn Plaza, New York, NY 10121-2298. Or contact your local bookstore.

This book is printed on recycled, acid-free paper containing a minimum of 50 percent recycled de-inked fiber.

For my mother, father, and brother
whom I credit with my success.

CONTENTS

ACKNOWLEDGMENTS

Steven van Zanen
Patrice Peyret
Patricia Longueuil
Stephanie De Labriolle
Sander Brouwer
William H. Dudley
Suzi Byer
Liz Attenburrow
Laura A. Matthews
Aram Krol
René Bruinsma
Çetin Çotuk
René Swankhuizen

WIRELESS MESSAGING DEMYSTIFIED

CHAPTER 1

An Introduction to the Wireless Industry

A great deal has happened in the past 10 years to completely transform the wireless industry. Many billions of dollars have been spent on infrastructure, and several generations of mobile phone technologies have come and (almost) gone. Billions more were spent merely on licenses to provide advanced wireless services in a variety of countries, and adoption rates have soared.

In the United States alone (according to recent statistics from Gartner Dataquest), 47 percent of the population owns a mobile phone. In Western Europe, that number now stands at an average of 75 percent, with adoption rates in Italy reaching as high as 88 percent.

The numbers are impressive and the ubiquity of mobile phones and wireless services is highly evident to anyone spending significant time in any major city throughout the globe. What is less obvious is the huge variety of tasks that those wireless handsets and wireless data devices are undertaking.

Introduction

A few years ago, there were really only two things that could be accomplished with the two major types of wireless communications devices available. A pager enabled you to either be notified when you have a voice message or, in later incarnations, enabled you to read who the message was from or read text messages. Mobile phones, meanwhile, were originally offered exclusively as devices for making and receiving calls.

Today, all of those lines have completely blurred. Today's wireless devices can send and receive e-mail, SMS messages, or even multimedia messages (including music, pictures, animated sequences, and even short bursts of video). They can also enable users to browse the Web, use the device as a speakerphone, and run a wide variety of computing applications (from personal information management to word processing to spreadsheets to games).

And it's a good thing that they can because that diversity of applications is being relied upon to drive the financial and technological future of the wireless industry. In an age where telecommunications companies are burdened with debt, handset manufacturers are seeing a slowing growth in sales, adoption rates are already high, and consolidation is under way throughout the industry, there is a huge demand for products and services that will drive revenues and margins for the industry and improve the *return on investment* (ROI) for users.

Enhanced networks enable new services (such as enhanced text messaging, web browsing, and e-mail) that drive the demand for the new handsets and wireless data devices that can take advantage of those services. New services are then introduced to take advantage of some of the untapped capabilities of the newly installed base of next-generation handsets and networks. In turn, revenues from sales of handsets and new services enable further upgrades to networks.

The Wireless Communications Paradigm Shift

Probably the biggest catalyst to change and development in the wireless sector throughout the past decade has been the growth in wireless data communications. Although many mobile phone users have appreciated improvements in the functionality of their handsets and the introduction of services such as conference calling, voicemail, call waiting, caller ID, and a huge variety of billing schemes that make voice calls more cost effective, the underlying key service (that is, enabling people to talk to one another wirelessly) has not changed fundamentally.

Voice calls are now clearer, coverage areas are much wider, and roaming agreements enable users to travel from country to country without an interruption of service. Essentially, however, voice services are still all about being able to make voice calls wherever and whenever you want.

That is why the entire wireless industry has undergone a massive paradigm shift to focus on the added-value delivery of wireless data services. Calling home on a slightly clearer line is nice, but being able to send and receive your office e-mail from your wireless handset can fundamentally change the way you work. Cheaper voice calls are good, but being able to instantly access the wireless Web to find out where you are and how to get where you want to go can provide real, measurable benefits to both individuals and corporations.

This shift in how and why wireless technology is used has been so profound that many analysts predict it will only be a few years before the amount of wireless data traffic eclipses voice traffic. That has already happened in the traditional wired network. And that trend, in and of itself, will be the catalyst for a whole range of other changes throughout the industry.

The first of those stems from the way wireless data traffic is charged. For example, charges for *Short Message Service* (SMS) messages (the pioneering wireless data technology that has become hugely popular in Europe and Asia throughout the last six years) have been billed in a variety of different ways.

They have been charged either purely on a per-message basis or on the basis of enabling users to send and receive a certain number of messages for free (in other words, built into the cost of the monthly airtime package charges) and then for a charge of *x* amount per message thereafter. In a number of markets, users were offered unlimited sending and receiving of SMS messages as an inducement to buy a mobile phone and sign up for an airtime package. Other carriers allowed free reception of SMS messages (opening the floodgates for SMS-based advertisers). Many carriers now also offer a variety of airtime packages, one or more of which will include sending and receiving an unlimited number of SMS messages.

The development of e-mail *gateways* for SMS, which enabled users to use SMS to send and receive e-mail to Internet-based e-mail systems, was also important to the growth in popularity of SMS (and also differentiated it further from voice messaging).

In markets where Internet access was expensive and users had to pay per-minute call charges for local calls to their local *Internet service provider* (ISP), SMS was also a way for many users to enjoy the benefits of electronic messaging at a low cost.

SMS messaging also gave users the ability to time shift conversations and undertake mobile conversations in situations where it would otherwise be difficult to do so. Users on a train or bus, for example, could swap SMS messages with friends and colleagues without being concerned about levels of background noise or being rude.

None of this is to say that SMS did not have drawbacks. To start with, it only offered a very limited number of characters per message, and messages had to be typed on a numeric keypad. But for early SMS adopters, this challenge was almost a badge of honor and the ability to send clever, yet coherent, messages with only a few characters became popular.

From an industry perspective, the key difference between this type of message-based wireless communication and voice communication was that there was no notion of an airtime charge for SMS based on the time of day or geography. No matter whom the message was being sent to, the charge was the same.

The messages were also delivered instantly and directly. Users did not have to dial in anywhere to pick up SMS messages. They were either notified by an audible tone, vibration, or the appearance of an onscreen alert. In fact, SMS messages effectively captured the mode of communication popularized a few years earlier by two-way pagers and brought that functionality within the mobile phone.

SMS was, of course, only the beginning of ubiquitous wireless data usage. Throughout this book, we will look in detail at how wireless messaging has evolved from its humble beginnings and the huge implications of the paradigm shift to wireless messaging as new services, upgraded networks, handsets, and wireless data devices are rolled out.

What Wireless Technology Means in Today's World

In today's world, wireless technology means many things to different users. That fact is due to the increasing popularity and variety of wireless data applications—the most popular of which is wireless messaging. This is in marked contrast to the days when wireless first came into common parlance as a noun. Halfway through the last century, "using on the wireless" meant listening to the radio.

Wireless described a device that could deliver only one experience of wireless—the delivery of sound. Through the evolution of wireless data communications, wireless now describes an entire industry that acts as a platform for delivering two-way communications across a variety of media, including sound, still images, full-motion video and live audio, computer-generated animation, and text-based content and messages.

Most of all, today's wireless solutions are all about providing a real ROI for both wireless providers and customers. Providers have made massive investments in infrastructure, spectrum licenses, marketing, handset subsidization, and research and development. Users have invested money in several generations of handsets, put forth effort in changing phone numbers and reconfiguring voicemail systems as providers evolved their offerings, and spent time learning and relearning the intricacies of using those handsets. They will also have spent innumerable hours trying to understand ever-changing calling plans, service offerings, and discount schemes.

More than anything, users of wireless services have invested their trust in the services themselves. They have trusted that they will enjoy widespread, reliable coverage, competitive prices, good customer support, and access to new voice and data services as they become available. Above all, they will have entrusted some aspects of operating their businesses and the day-to-day workings of their personal lives to wireless services.

Users enjoy a measurable ROI in wireless voice and data technology in circumstances such as the following:

- The real-estate agent who is able to follow through a tough negotiation while on the road
- The engineer who can send urgent reports from the field to ensure that the right materials for a big project arrive at the right place at the right time
- The corporation that knows it can instantly send messages to all its field sales reps throughout the region with confidence that they will all see the messages right away and respond appropriately

These are but a few examples of the many ways in which measurable ROI can be achieved for businesses that choose a wireless solution. And with the rise of wireless data within the paradigm shift described earlier, many more businesses will be seeking and finding this ROI as the cost of wireless data services declines and the range of such services increases.

The Reality of Wireless Hype

It has been said that modesty is perhaps one of life's most overrated virtues. But recent history suggests that a little modesty and humility would have gone a long way in the wireless industry over the past decade.

One of the biggest roadblocks to realizing the full potential of wireless voice and data services has been created by the industry itself—an expectation of massive benefits that are completely outside of what the industry is currently capable of delivering (and, in some of the worst cases of wireless hype, completely beyond what wireless services will ever be able to deliver).

This expectation helped to fuel the wireless bubble that artificially inflated the share price of companies offering wireless services and caused many consumers and business users alike to be deeply disappointed in their initial encounters with *next-generation* wireless services.

It's probably fair to suggest that the zenith (or nadir, depending on your point of view) of wireless hype came between February 28 and March 1, 2000 at the *Cellular Telecommunications Industry Association* (CTIA) conference in New Orleans.

All the glittering digirati of the dot com and wireless boom were present from Microsoft co-founder Bill Gates (also known by his not-so-secret identity of The World's Richest Man) to Amazon.com CEO Jeff Bezos (fresh from an appearance on the cover of *Time Magazine*) to AOL CEO Steve Case (fresh from buying the whole of Time Warner with AOL stock).

As companies such as *Research in Motion* (RIM), Sierra Wireless, Motorola, Nokia, Bell Atlantic, and many others stepped up to the podium at CTIA 2000, their share prices hit all-time highs and the world's expectations of wireless services went through the roof along with them. The hype came thick and fast. Here's just a sample of what was said:

> There's no doubt this is an industry sitting on the verge of unbelievable opportunity, not just connection to full-screen PC devices, but making that seamless with the other devices, defining new types of communication, the location-based services, the e-everything from photos to books to music, making sure that there's a way to reach into the secure information. It's not just the data on the Internet; it's also a lot of information like your SAP applications or other things. You want notifications coming out of that world, and yet there's a lot of requirements to make sure you integrate and the security is preserved. So, the devices themselves will be surprising us all with the great things they can do. There will be a lot of magic software there. The networks are going to be surprising us. And we're very, very excited to be working with all of you to seize this opportunity.
>
> —Bill Gates at CTIA 2000

Although Gates was less full of hyperbole than many, his very appearance at an event that was pumping the promise of wireless for all it was worth gave credence to many of the more extreme claims of wireless benefit. The hype was juxtaposed against the comments of a real wireless pioneer, Apollo 11 astronaut Neil Armstrong, the first man to set foot on the moon, whose mission involved sending wireless voice and data transmissions from the surface of the moon. Ironically, he had one of the most down-to-earth speeches of the entire event.

In the wake of that hype fest, many wireless companies went into a precipitous share price slide and customer credibility crisis from which some never recovered.

Wireless and Hype Go Hand in Hand It's hard to blame wireless companies for hyping their products and services or analysts for making extremely rosy predictions for the future of wireless initiatives such as the many overblown and overpromoted *mobile commerce* (m-commerce) services prelaunched or preannounced in 2000 and 2001.

The reason that wireless technology inspires such hype is that it *is* possible to offer a vast number of services wirelessly. The question, however, is no longer about what is possible, but rather about what is economic, profitable, and sensible.

Just as it was possible for thousands of venture capital-based, content-based web sites to offer free news, reviews, streaming video, and music for a while, it was possible for wireless providers with inflated stock prices and bank accounts to provide massive subsidies for all kinds of innovative wireless services in the absence of any demand for such services for a while. But, in the end, those services have to be services that users will pay for on a sustainable, profitable basis.

Technology made it possible to envision all kinds of possible wireless services that could be offered and hyped, but it has become very important for all players in the wireless industry to sift through the possibilities until all that remains are those services that are sensible, profitable, and achievable (yet still innovative and desirable).

A Good Hype Filter The best possible filter for wireless hype is very simple. When looking at any proposed wireless service, just ask one question: Is this a technology looking for a solution, or a technological solution to an existing need?

In other words, it is much harder to sell people a solution to a problem that they don't know they have than it is to sell them a solution to a problem that is currently causing them a good deal of grief. In the first instance, the solution provider has to go looking for customers. In the second, customers will come looking for the solution provider.

So when wireless service providers started promising future wireless data services that were completely uneconomical and for which there was little or no demand, it should have come as no surprise that the services never made it out of the starting blocks.

Probably the most shining example of that is wireless videoconferencing. Although there is no doubt that a small, well-defined niche may exist for wireless videoconferencing, there is no concrete evidence yet that anything resembling a mass market exists for this technology, despite the fact that with the speed of available wireless networks in a number of markets, it is entirely possible to offer such a service.

The Wireless Playing Field

The wireless market is not a simple world of customers and suppliers. It is a richly textured environment that includes a number of additional players, such as handset manufacturers, government regulators, and strange beasts known as *mobile virtual network operators* (MNVOs). The key players are discussed in the following sections.

Governments

Governments play a far bigger role in the wireless sector than it might appear to the casual observer. Firstly, and perhaps most importantly, governments hold the power to license spectrum to the wireless operator/carriers of the world. Different countries have different processes for doing so, and the licensing process may have as much to do with national political issues as it does with the merit of a particular bidder's proposal.

Although spectrum allocation is perhaps not as big an issue now as it was a few years ago (as all the key spectrum auctions for next-generation wireless services have been held), the impact of governments' role in this will likely still be felt for a long time to come. Many governments saw the spectrum auctions of the late 1990s as a way to make a fast buck off the skyrocketing wireless industry. Wireless companies ended up spending a great deal of the money they had raised on the then-booming equities markets to buy spectrum licenses. Even worse, a number of them took on huge debt loads to pay for the licenses, which has come back to haunt them.

National, regional, and local governments have also played a part in the rollout of infrastructure to support wireless networks. In many markets, governments needed to grant permission for the construction of strategically located base stations, transmitter towers, and repeaters. Failure to receive such permission in a given area would seriously impact the viability for wireless service in that region.

Government health and safety regulations have also had an impact on the wireless industry. Not only is there ongoing, government-funded research into the potential health hazards produced by wireless handsets, but there are also government-mandated services (such as *Enhanced 911* [E911]) that must be built into networks and supported by handsets.

The U.S. *Federal Communications Commission* (FCC), for example, has mandated the establishment of wireless E911. Its E911 rules seek to "improve the effectiveness and reliability of wireless 911 service by providing 911 dispatchers with additional information on wireless 911 calls."

The wireless E911 program is divided into two parts—Phase I and Phase II. Phase I requires carriers, upon appropriate request by a local *Public Safety Answering Point* (PSAP), to report the telephone number of a wireless 911 caller and the location of the antenna that received the call. Phase II requires wireless carriers to provide far more precise location information, within 50 to 100 meters in most cases.

The deployment of E911 requires the development of new technologies and upgrades to local 911 PSAPs as well as coordination among public safety agencies, wireless carriers, technology vendors, equipment manufacturers, and local wireline carriers. The FCC established a four-year rollout schedule for Phase II, beginning October 1, 2001 and completing by December 31, 2005.

So governments play a role in a wide variety of areas in the wireless industry. In some countries, governments may also be significant shareholders in wireless operators as part of a legacy investment in a monopoly wireline phone service provider.

Operators/Carriers

This group is at the core of the wireless storm and currently finds itself in the midst of a broad evolution. Mobile phone service operators have typically not only owned and operated a wireless network, but they have also typically handled their own marketing, distribution, sales (through retail partners and/or wholly owned retail subsidiaries), support, billing, and service.

The business model for operators—many of which started life as extensions of an existing wireline telephone monopoly or oligopoly in a given region—has been that of the wireline carrier. Over time, a number of operators have looked very hard at this structure to see whether it makes sense for the medium and long term.

Some operators have divested themselves of billing and outsourced that function or spun out the division that provides it. Others have done the same with retail operations or marketing. Some have even gone as far as whittling down their basic function to merely operating the network and receiving fees from any company that wants to offer wireless services to businesses or consumers on that network (with former divisions of the operator's parent company usually being the biggest customer).

Operators currently face many challenging issues from reevaluating their overall structures to managing subscriber churn to juggling debt to dealing with falling levels of shareholder satisfaction. And they are doing so in an environment where they are expected to roll out next-generation services, maintain healthy profit margins, and build on rates of growth that must seem impossible to sustain.

Handset Manufacturers

This group has probably the most straightforward role to play in the industry. Handset manufacturers create the devices that will leverage the value of the networks run by the network operators. They also have a highly symbiotic relationship with them. A good handset manufacturer can be a huge boost to the fortunes of a wireless network operator, helping to drive business and consumer demand for the services of the operators.

Perhaps a good analogy for the role of the handset manufacturer in the wireless industry is the automotive world. In many ways, cars are merely the vessels that create a global system of gasoline delivery and consumption. For the average consumer, the whole experience is about the car. For the industry as a whole, the consumption of gasoline is a much greater source of daily revenue.

In the wireless industry, consumers may think about handsets as the public face of the services they use. They may associate their whole wireless experience with the appearance and features of the handset. Yet the vast majority of the revenue earned for the industry during the lifetime of that handset will go to the operator. The handset merely provides a way of delivering the wireless services to the consumer.

However, there is a strong symbiosis between handset manufacturers and operators. Handset manufacturers only make money when businesses or consumers buy handsets. It is in their interest to ensure that handset users change their handsets every few years or sooner.

As competitive pressures constantly drive down the price of airtime and added-value services (for example, voicemail goes from being a monthly added-value item to being part of a monthly bundle necessary to compete with an arch rival's monthly bundle), operators are constantly in search of new, higher-margin services that they can roll out to customers. As a result, they need new handsets to deliver those services.

There would be little point, for example, in rolling out a service to corporate clients that enables them to securely access their Microsoft Exchange or Lotus Notes e-mail servers from their handsets if this was not a feature that handset manufacturers supported. However, if such a feature is going to help drive handset sales, manufacturers will most definitely be lining up to provide such features.

Then there is the issue of handset subsidies. In order to sell consumers or business users on long-term (two- or three-year) contracts for wireless services, operators often provide handsets at highly discounted prices (hence the popular $0 mobile phone). The cost of the mobile phone is actually factored into the price of the contract.

What all of this means for manufacturers, however, is that they end up being very closely allied with operators as this heavily subsidized and discounted retail handset pricing ensures that consumers largely buy handsets with service contracts rather than paying full retail price for them. As a result, handsets are rarely sold to consumers without a service contract. When they are, consumers pay much more up front, which they may be willing to do if they have a network operator they are happy with and the transition is merely a matter of swapping *Subscriber Identity Module* (SIM) cards from the old handset to the new one.

The rise of prepaid (or pay-as-you-go) services in a number of markets has served to change this relationship somewhat, as retailers must collect money for the handset up front since they have no guarantees about how much and what kind of wireless services users will consume when they leave the shop.

The initial promise of prepaid services was that they could lower the bar for users to start consumer wireless services, while operators would benefit because prepaid users would be tied to them through a desire not to change handsets, service providers, or voicemail systems. Operators also hoped for degree of *breakage* (prepaid users buying airtime cards and not using the full value of the cards before they expired, thereby enabling operators to make greater profit on the card).

The reality, however, is that although prepaid services have been a bonanza for consumers (particularly students) and have done a great deal to help pioneer all kinds of innovative billing and service delivery models, they have not always been highly profitable for either handset manufacturers or carriers. In a number of markets, handset manufacturers have discovered that prepaid customers tend to buy the cheapest handset they can. Operators have discovered that users will either exploit whatever services are offered to prepaid users for free (some have offered incoming calls or text messaging for free), whereas others will use a prepaid handset as an emergency phone that sits in a glove compartment or a kitchen drawer (along with the flashlight, matches, and first-aid kit).

The role of handset manufacturers is not merely limited to that of being a supplier to network operators. Manufacturers also participate in discussions with global, national, and regional standards bodies and work to develop products that are appropriate for them.

Consumers

Consumers are in an odd position in the wireless market. On the one hand, they are being asked to expect less from their wireless services (such as the vast array of free content-based services initially made available to Internet-enabled handsets). On the other hand, they are always being promised more (more messaging-based services, improved handsets, higher data speeds, and so on). Consumers have undergone the experience of being at the heart of a great deal of market experimentation, which has pushed the all-important consumer expectations all over the map.

The pricing of handsets is a prime example of that. By heavily subsidizing the price of handsets for consumers by building the cost of them into two- or three-year service contracts and portraying the cost of the handset as being either free or close to it, carriers did a great job of creating an expectation among consumers that they should pay little or nothing for handsets.

This situation required handset manufacturers to try to make higher margins on after-sale accessory products (such as batteries, in-car kits, data kits, face plates, headsets, and so on). It also made life difficult for carriers when they later went back to the consumer market with prepaid offerings that actually required the consumer to pay more of the real cost of the handset—something that proved a challenge since consumers were used to handsets being inexpensive.

Similarly, the overpromise of services in the initial flush of Internet-enabled handsets also led to a good deal of consumer disappointment. Once they had become used to full-color displays, graphics, animation, and even streaming audio and video, the slow delivery of a few lines of text on a tiny green and black screen seemed to pale by comparison.

MNVOs

MNVOs are a product of operators rethinking how they should participate in the wireless sector, and what roles branding and scalability could play in their success or failure. The first characteristic of an MNVO is that an MNVO does not own spectrum—it offers wireless services by paying for access to a wireless infrastructure that is commonly owned by a traditional operator.

MNVOs can be an exercise in brand extension (which you could argue was the principal motivation behind the U.K.'s Virgin Mobile), or they can be a way of enabling traditional carriers to split into a piece that actually handles subscribers, billing, marketing, and customer-facing activities while another part of the business becomes largely a spectrum infrastructure operation (that may or may not have the ability to provide the company's spectrum to other competitors).

MNVOs are not entirely unlike the many long-distance competitors that sprang up throughout the 1990s by reselling long-distance services to consumers (after buying those services in bulk from traditional carriers).

Some MNVOs also came into existence as a way of creating international wireless services for businesses and consumers by buying the rights to resell services from a number of national providers—either by stitching together a branded service or through strategic roaming agreements.

The Value Proposition of Wireless Messaging

The value proposition behind wireless messaging is key to the continued and future success of the wireless industry. Just as e-mail has been a massive driver for the growth in Internet adoption and sales of telecommunications services and computer hardware in the past eight years, wireless messaging is proving to be the engine of growth for the wireless industry.

The value proposition of wireless messaging is not only the ability for consumers and corporate users to send and receive rich, meaningful messages anytime and anywhere. It also lies in the ability to manage and optimize that experience, so that users get the messages they really want when they want them and are able to get important messages to others when it is vital to do so (and authenticate the delivery and acceptance of those messages).

For anyone who has ever spent time playing telephone tag in order to move a business process forward, the notion of being able to reliably have instant communication with a key colleague or partner will have immediate value and resonance. Wireless messaging provides a cost-effective, reliable, and universal way for this to happen.

Wireless messaging further provides a way for the complex web of messages that make up daily working life in the twenty-first century to be better managed. Voicemail, e-mail, text messages, and fax messages can now all be part of a well-managed wireless messaging service. The computing power built into today's wireless handsets, with greater intelligence in the networks that serve them, provide a variety of different ways in which users can achieve value and ROI by prioritizing messages and helping people make more effective use of the massive amount of information that comes at them everyday.

Today's Business Drivers for Wireless Messaging

If you had a single phrase that identifies a strategic imperative for business today, it would be the phrase "do more with less." Businesses are being required to do more with fewer people, more limited access to capital markets, a tougher lending environment, and a shorter leash from regulators and investors alike.

In order to meet the business imperatives suggested by this phrase, businesses have to leverage existing investments in both their IT and telecommunications infrastructures. "Rip and replace" is no longer an option for many businesses. That means they need to get better intelligence from the information they collect from transactions, act on the implications of that information better, and communicate more effectively with suppliers, partners, distributors, and customers.

For any new product or service offered to corporate users today, there must be business drivers that meet one or more of the following criteria:

- It saves the company time or money.
- It helps the company make money.
- It is a requirement for doing business.

The business drivers for wireless messaging currently fall into both of the first two categories, and the wireless messaging industry is working hard to find ways to put their offerings in the third.

Saving Money

Here's how those business drivers break down.

- It provides time-efficient delivery of messages wherever the recipient happens to be. Wireless messages are delivered to the receiving device almost instantaneously. In terms of the timeliness of location-independent message delivery, there is little that can touch wireless messaging systems. Fax systems typically require use of a landline, voicemail (even on mobile phones) requires that users dial in to receive it, and traditional e-mail requires being close to a fixed Internet connection. If time is money and wireless messaging saves time in delivery messages, then it saves money.
- *It reduces airtime and long-distance charges.* Wireless messages are typically composed offline and a connection to the network is only made for the brief split seconds that it takes to send the text over the network. Pricing is also not traditionally based on the time required to send or receive the message; instead, it is based on per-message charges. The sending and receiving of those messages is also usually independent of location, eliminating the traditional notions of long-distance or roaming charges associated with wireless voice communications.
- *It encourages quick, efficient queries and responses.* It is true to say that wireless devices (at least those that are currently available) are not ideal for composing or reading long, involved messages. In fact, they encourage an economic use of time in sending messages to others and creating responses. When used effectively, this can lead to being able to reduce a lot of day-to-day communication to the essential nuggets of information to be exchanged—something that social convention would typically not allow in a voice conversation or even a more extensive e-mail exchange.
- *It is very direct.* Companies can configure wireless devices so that users only receive messages that are highly particular to

their work priorities, giving them a tool to sift through the mass of messages that might hit them in a given day and providing employees with a way to achieve greater focus.

- *It distributes response time.* By allowing mobile employees to respond wirelessly, they can more effectively use time that might not otherwise be available to them for fielding written queries. Without wireless solutions, those mobile employees may have to "shoe horn" their e-mail responses into a few hours at the beginning or end of the day, which not only delays responses, but it also puts undue stress and pressure on those employees.

Making Money

- *Sales staff can respond to customers in a more timely fashion.* By being able to receive customer queries wirelessly and respond to them in real time (or as near to real time as they want), mobile sales staff are equipped to more effectively follow up with customers and gain the maximum yield from each sales opportunity.
- *Gain greater access to sales support information.* By being able to instantly get answers about a given client or sales prospect through a wireless messaging system, sales staff can be more informed, look smarter, wirelessly send and create quotes, and more quickly close sales with clients.
- *Evaluate opportunities.* In many business opportunities, you snooze, you lose. If you can't get quick access to the information you need to evaluate a potential business opportunity quickly, you can be sure that your competitor will find a way. In a world of fast-changing and volatile markets, timing can be the difference between making and losing money.
- *Provide effective management.* Managing everything from cash flow to employees is hard when you are working with old information. Wireless messaging provides a way for managers of companies large and small to have a delivery mechanism with them at all times that enables the delivery of rich, meaningful information that helps them make decisions that can make money for the companies they manage.

A Requirement for Doing Business

It's not there yet, but things are moving in that direction. In some industries (such as equities and financial services sector), wireless messaging has become so important that data-enabled wireless handhelds or dedicated wireless messaging devices (such as the RIM Blackberry) have now come very close to being vital for doing business.

As the bar of customer service is raised by those organizations that serve their customers better through strategic use of wireless messaging, there will be increasing competitive pressures for companies to designate wireless messaging devices and services as a requirement for doing business.

When that happens, wireless messaging moves to the point of being an assumed cost for the operation of any effective business. Just as no one today would conceive of operating a business without a telephone (or e-mail, for that matter) and would not spend time evaluating the cost justification for having a telephone, this would be the case for wireless messaging once it reaches that stage.

The Wireless Value Chain: Who Supplies What to Whom?

Figure 1-1 is a diagram, originally developed in British Columbia by the Canadian Institute for Market Intelligence, used to show the participation of local companies in the wireless industry. It does a good job of explaining the key elements of the wireless value chain. It may differ slightly in operation in different countries, but it captures the underlying structure of the value chain quite effectively.

As you can see, the value chain starts with the component manufacturers for both wireless devices (such as handsets) and the wireless network infrastructure. Without them, there would be no networks or any devices to use on the networks. They comprise the equipment component of the value chain.

Then come the network components of the value chain, such as wireless data carrier/operators and the MVNOs, and the enabling

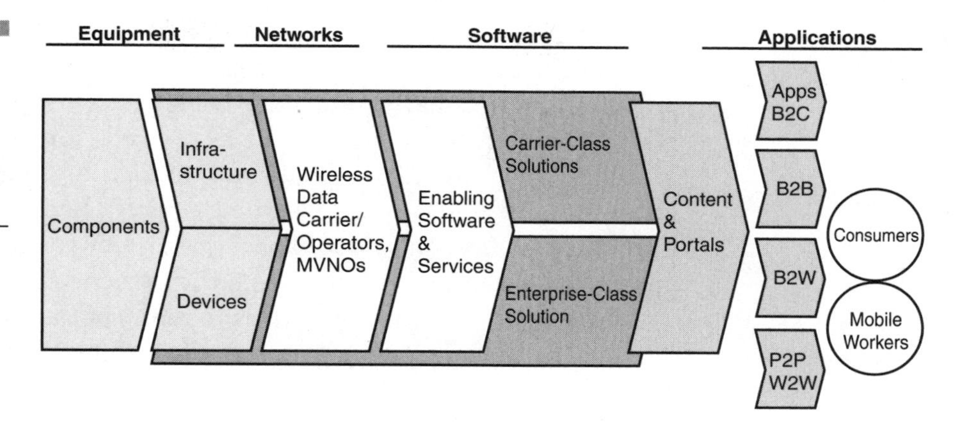

Figure 1-1 Wireless industry participants (Source: Canadian Institute for Market Intelligence)

software and services that they use to operate their respective businesses. Taken together, these groups are involved in maintaining, marketing, and enhancing the network.

Next in the value chain comes the software that really makes the paradigm shift in wireless use. This wireless application software is typically divided into two classes—software to serve the needs of the carriers (and underpins the solutions that they can then offer to consumers) and software that underpins the requirements of the carriers' respective corporate customers (the enterprise-class solutions).

Finally, there are the applications themselves, which may actually be marketed as services or elements of service bundles by the carriers. These include the operation of content wireless services and portals, and business-to-consumer, business-to-business, web-to-wireless, peer-to-peer (such as text messaging), and business-to-wireless services.

At the far end of the value chain, of course, are the actual users of these services—consumers and mobile workers—who will typically use some combination of the applications bundled as services.

Why GSM Operators Are Ahead

Operators of networks that conform to the *Global Standard for Mobile* (GSM) have gained a jump on the rest of the wireless industry in providing strong business drivers for wireless messaging and in the general use of wireless services.

The biggest business driver was simply the fact that GSM was a standard in so many different countries, enabling users to roam from one country to another and still be able to carry the same mobile phone number, handset, and set of services. In the European communities, this provides a major advantage as mobile workers and consumers travel a lot between countries. Having a common standard ensures that they would be able to enjoy the full benefit of their wireless messaging services while they traveled.

As far as pricing goes, operators in many of the countries that offer GSM services were smart as well. They offered services on a basis that was highly familiar to business telecom consumers in those countries—through *Calling Party Pays* (CPP) billing. Businesses and consumers throughout Europe, for example, were used to the notion that they would pay for every outgoing call they made and that all incoming calls would be paid by the party initiating the call. And this has long been the case, regardless of where they were calling. Unlike North America, where per-minute call charges for landline telephony were not levied on local calls, most European countries have always charged on a per-minute and CPP basis for all calls.

This was in sharp contrast to the North American experience, where per-minute charges were always associated with premium services (such as long-distance or cellular services). In the case of cellular voice calling, North American wireless operators would typically levy a per-minute charge on both the caller and receiver.

The result of this combination of historical and pricing model difference was that while European users of mobile phone services were encouraged to give out their mobile phone numbers (because they knew that customers or potential clients could call them at no additional expense to themselves), North American users were encouraged to be sparing in their distribution of their mobile phone numbers (because they would be on the hook for the cost of the airtime, and potentially long-distance charges, when someone called them).

That fact was also coupled with the cultural difference that North Americans were not used to paying per-minute call charges as part of their day-to-day business operations (other than for long-distance services, for which many large customers would buy dedicated lines to eliminate or lower such charges). As a result, their use of mobile services was initially quite tentative for fear of running up high bills. Europeans, meanwhile, had been working with per-minute call charge schemes for years and were not in the least concerned by them.

The Value of Prepaid Wireless

Prepaid wireless services also had a lot to do with the initial success of wireless messaging in countries that provided GSM services. Again, the services were provided on the basis that the caller pays so that consumers could buy prepaid mobile phones for members of their family and ensure that they could use the phone to reach their loved ones at all times—no matter how little airtime was left on their prepaid calling plan.

The same was true for mobile employees in small businesses, who could be equipped with prepaid mobile phones and a certain number of phone cards per month. If they wanted more, they could buy them. But it gave the business a way of controlling costs while also ensuring that it could stay in touch with employees.

Again, pricing models had a lot to do with the way that wireless messaging was used. In some cases, text messaging was a free service for prepaid users, provided they had at least some airtime left on their mobile phones. In other cases, text messages could be sent at a far lower cost than calling someone as text messages were charged on a flat per-message basis, while calls would depend on where you were calling someone and what time of day it was. Students, for example, could text message their parents and then have the parents call them, thereby making a call that cost almost nothing for the student (as the text message would have cost a few pennies and the incoming call would be free).

Prepaid wireless also did a great job of educating users about the value of mobile phones and wireless data services, so that they were better candidates to be sold a higher level of service and services through using a postpaid account.

In North America, prepaid did not work so well—again partially due to the less attractive billing and partially due to a cultural discomfort with the idea of paying per-minute call charges.

Introduction to Acronyms and Industry Terms

In order to provide the most consistent and broadly agreed-upon definitions of the standards and acronyms currently in use by the wireless industry, we quote some of the alphabetically organized descriptions given by the CTIA at its web site www.wow-com.com:

Air interface The standard operating system of a wireless network; technologies include AMPS, TDMA, CDMA, and GSM.

AMPS (Advanced Mobile Phone Service) The term used by AT&T's Bell Laboratories (prior to the breakup of the Bell System in 1984) to refer to its cellular technology. The AMPS standard has been the foundation for the industry in the United States, although it has been modified in recent years. *AMPS compatible* means equipment designed to work with most cellular telephones.

APCO (Association of Public Safety Communications Officials-International) A trade group headquartered in South Daytona, Florida, representing law enforcement, fire, emergency services, and other public safety agency dispatchers and communications employees.

CALEA (Communications Assistance to Law Enforcement Act) A 1994 law granting law enforcement agencies the ability to wiretap new digital networks and requiring wireless and wireline carriers to enable eavesdropping equipment use in digital networks.

CDMA (Code Division Multiple Access) A spread spectrum approach to digital transmission. With CDMA, each conversation is digitized and then tagged with a code. The mobile phone is then instructed to decipher only a particular code to pluck the right conversation off the air. The process can be compared in some ways to an English-speaking person picking out in a crowded room of French speakers the only other person who is speaking English.

CDPD (Cellular Digital Packet Data) An enhanced system overlay for transmitting and receiving data over cellular networks. Technology that enables data files to be broken into a number of packets and sent along idle channels of existing cellular voice networks.

Cell The basic geographic unit of a cellular system. Also, the basis for the generic industry term *cellular*. A city or county is divided into smaller cells, each of which is equipped with a low-powered radio transmitter/receiver. The cells can vary in size depending upon terrain, capacity demands, and so on. By controlling the transmission power, the *radio frequencies* (RFs) assigned to one cell can be limited to the boundaries of that cell. When a wireless phone moves from one cell toward another, a computer at the *Mobile Telephone Switching Office* (MTSO) monitors the movement and at the proper time, transfers or hands off the phone call to the new cell and another RF. The handoff is performed so quickly that it's not noticeable to the callers.

CMRS (Commercial Mobile Radio Service) An FCC designation for any carrier or licensee whose wireless network is connected to the PSTN and/or is operated for profit.

Cost recovery Reimbursement to CMRS providers of both recurring and nonrecurring costs associated with any services, operation, administration, or maintenance of wireless E911 service. Costs include, but are not limited to, the costs of design, development, upgrades, equipment, software, and other expenses associated with the implementation of wireless E911 service.

ESMR (Enhanced Specialized Mobile Radio) Digital SMR networks, usually referring to Nextel Communications Inc., which provide dispatch, voice, messaging, and data services.

ESN (electronic serial number) The unique identification number embedded in a wireless phone by the manufacturer. Each time a call is placed, the ESN is automatically transmitted to the base station so the wireless carrier's mobile switching office can check the call's validity. The ESN cannot be altered in the field. The ESN differs from the MIN, which is the wireless carrier's identifier for a phone in the network. MINs and ESNs can be electronically checked to help prevent fraud.

FCC (Federal Communications Commission) The government agency responsible for regulating telecommunications in the United States.

FWA (fixed wireless access) Also known as WLL.

GPRS (General Packet Radio Service) A GSM data transmission technique that does not set up a continuous channel from a portable terminal for the transmission and reception of data, but transmits and receives data in packets. It makes very efficient use of available radio spectrum, and users pay only for the volume of data sent and received.

GPS (Global Positioning System) A series of 24 geosynchronous satellites that continuously transmit their position. Used in personal tracking, navigation, and automatic vehicle location technologies.

GSM (Global System for Mobile) communications A digital cellular or PCS network used throughout the world.

iDEN (Integrated Digital Enhanced Network) A Motorola Inc. ESMR network technology that combines two-way radio, telephone, text messaging, and data transmission into one network.

ILEC (Incumbent Local Exchange Carrier) The historic local phone service provider in a market, often a former Bell company. Distinct from CLECs, new market entrants.

IMT-2000 The International Telecommunication Union's name for the new 3G global standard for mobile telecommunications.

IS (Interim Standard) A designation of the *American National Standards Institute* (ANSI), usually followed by a number, that refers to an accepted industry protocol, for example, IS-95, IS-136, and IS-54.

IS-661 North American standard for 1.9 GHz wireless spread spectrum RF access technology developed by Omnipoint Corp. IS-661, for which Omnipoint was awarded a pioneer's preference license for the New York City market, is based on a composite of CDMA and TDMA technologies. The company says IS-661 reduces infrastructure costs and allows higher data speeds than mainstream GSM or TDMA platforms.

ISDN (Integrated Services Digital Network) An advanced, high-capacity wireline technology used for high-speed data transfer.

ITU (International Telecommunication Union) An agency of the United Nations, headquartered in Geneva, that furthers the development of telecommunications services worldwide and oversees global allocation of spectrum for future uses.

MIN (mobile identification number) Uniquely identifies a mobile unit within a wireless carrier's network. The MIN often can be dialed from other wireless or wireline networks. The number differs from the ESN, which is the unit number assigned by a phone manufacturer. MINs and ESNs can be electronically checked to help prevent fraud.

NAM (Number Assignment Module) The NAM is the electronic memory in the wireless phone that stores the telephone number and an ESN.

NAMPS (Narrowband Advanced Mobile Phone System) NAMPS combines cellular voice processing with digital signaling, increasing the capacity of AMPS systems and adding functionality.

Narrowband PCS The next generation of paging networks, including two-way, acknowledgment, and wireless answering machine paging.

PCS (personal communications services) A two-way, 1,900 MHz digital voice, messaging, and data service designed as the *second generation* (2G) of cellular.

PDA (personal digital assistant) A portable computing device capable of transmitting data. These devices make possible services such as paging, data messaging, e-mail, computing, facsimile, date book, and other information handling capabilities.

PIN (personal identification number) A code used by a mobile telephone number in conjunction with an SIM card to complete a call.

POPs (persons of population) This term is used to designate the number of potential subscribers in a market.

PSTN (Public Switched Telephone Network) The worldwide voice telephone system, also called the Bell System in the United States.

PSAP (Public Safety Answering Point) The dispatch office that receives 911 calls from the public. A PSAP may be local fire or police department, an ambulance service, or a regional office covering all services.

PUC (Public Utility Commission) The general name for the state regulatory body charged with regulating utilities including telecommunications.

RBOC (Regional Bell Operating Company) The list of such companies includes Bell Atlantic, U.S. West, Ameritech, Southwestern Bell, and BellSouth.

SMR (Specialized Mobile Radio) A dispatch radio and interconnect service for businesses. Covers frequencies in the 220, 800, and 900 MHz bands.

SMS (Short Messaging Service) Electronic messages on a wireless network.

Spectrum allocation Federal government designation of a range of frequencies for a category of use or uses. For example, the FCC allocated the 1,900 MHz band for PCS. Allocation, typically accomplished in years-long FCC proceedings, tracks new technology development. However, the FCC can shift existing allocations to accommodate changes in spectrum demand. As an example, some *Ultra High Frequency* (UHF) television channels have been reallocated to public safety.

Spectrum assignment Federal government authorization for use of specific frequencies or frequency pairs within a given allocation, usually stated at a geographic location(s). Mobile communications authorizations are typically granted to private users, such as oil companies, or to common carriers, such as cellular and paging operators. Spectrum auctions and/or frequency coordination processes, which consider potential interference to existing users, may apply.

TDMA (Time Division Multiple Access) A method of digital wireless communications transmission enabling a large number of users to access (in sequence) a single RF channel without interference by allocating unique time slots to each user within each channel.

Third generation (3G) A new standard that promises to offer increased capacity and high-speed data applications up to 2 Mbps. It also will integrate pico-, micro- and macrocellular technology and allow global roaming. It is the next generation of wireless technology beyond PCS. The World Administrative Radio Conference assigned 230 MHz of spectrum at 2 GHz for multimedia 3G networks. These networks must be able to transmit wireless data at 144 Kbps at mobile user speeds, 384 Kbps at pedestrian user speeds, and 2 Mbps in fixed locations.

Tri-mode handset Phones that work on three frequencies, typically using 1,900 MHz, 800 MHz digital, or reverting to 800 MHz analog cellular when digital is not available.

ULS (Universal Licensing System) The new Wireless Telecommunications Bureau program under which electronic filing of license applications and reports of changes to licenses creates a database that can be accessed remotely for searches. Using ULS, for example, the user can learn all the specialized mobile radio licenses in a given region.

UMTS (Universal Mobile Telecommunications System) Europe's approach to standardization for 3G cellular systems.

WAP (Wireless Application Protocol) A de facto worldwide standard for providing Internet communications and advanced telephony services on digital mobile phones, pagers, PDAs, and other wireless terminals.

W-CDMA (Wideband Code Division Multiple Access) The 3G standard offered to the ITU by GSM proponents.

WCS (wireless communications services) Frequencies in the 2.3 GHz band designated for general fixed wireless use.

WIN (Wireless Intelligent Network) The architecture of the wireless switched network that enables carriers to provide enhanced and customized services for mobile telephones.

Wireless Using the RF spectrum for transmitting and receiving voice, data, and video signals for communications.

Wireless Internet An RF-based service that provides access Internet e-mail and/or the World Wide Web.

Wireless IP The packet data protocol standard for sending wireless data over the Internet.

Wireless IT (wireless information technology) The monitoring, manipulating, and troubleshooting of computer equipment through a wireless network.

WLAN (wireless local area network) Using RF technology, WLANs transmit and receive data over the air, minimizing the need for wired connections. Thus, WLANs combine data connectivity with user mobility. WLANs are essentially networks that allow the transmission of data and the ability to share resources, such as printers, without the need to physically connect each node, or computer, with wires. WLANs offer the productivity, convenience, and cost advantages over traditional wired networks.

WPBX (wireless private branch exchange) Equipment that enables employees or customers within a building or limited area to use wireless handsets connected to an office's PBX system. WPBX systems, for example, include a wireless handset that is programmed to ring simultaneously with the desk phone.

WLL (wireless local loop) WLL is a system that connects subscribers to the PSTN using wireless technology coupled with line interfaces and other circuitry to complete the last mile between the customer premise and the exchange equipment. Wireless systems can often be installed in far less time and at lower cost than traditional wired systems.

xDSL Designation for digital subscriber line technology enabling simultaneous two-way transmission of voice and high-speed data over ordinary copper phone lines.

WAP and DoCoMo

With some understanding of the key terminology and players in the wireless messaging sector, it is important to consider which of these technologies is actually delivering on the promise of wireless messaging and which have set it back. There is a strong argument to suggest that WAP has actually set back the case of wireless messaging.

WAP was marketed as a way of bringing the Internet to mobile phones, but it was actually a way of bringing a very limited content and applications subset of the Internet and the Web to the wireless world. It is a classic case of overpromise and underdelivery, and stands in stark contrast to the success of DoCoMo's i-mode service in Japan. WAP was a case of shooting for the lowest possible technological and telecommunications denominator—DoCoMo shot for the highest, developing a structure that could be used to exploit the performance and capability of 3G standards and many multimedia components.

In a recent speech at the Accenture Global Convergence Forum on Monte Carlo, a senior executive of NTT DoCoMo USA offered a perspective on this issue. He said that there are now more than 73 million subscribers to cellular telephone services in Japan, which also has almost 50 million wireless access Internet subscribers (see Figure 1-2).

In fact, the number of cellular phone subscribers and wireless access Internet subscribers both overtook the total of fixed-line telephone subscribers in 2000. He added that wireless is moving much quicker than fixed-line business. Whereas it took five years for the Internet to reach 10 percent market penetration, wireless Internet took only one year.

Potential expansion opportunities exist, he said, in moving from voice to nonvoice or multimedia services, from personal to ubiquitous applications for any mobile entity, and from domestic to international services.

Figure 1-2 Wireline vs. wireless growth in Japan compared (Source: NTT D.C.M.)

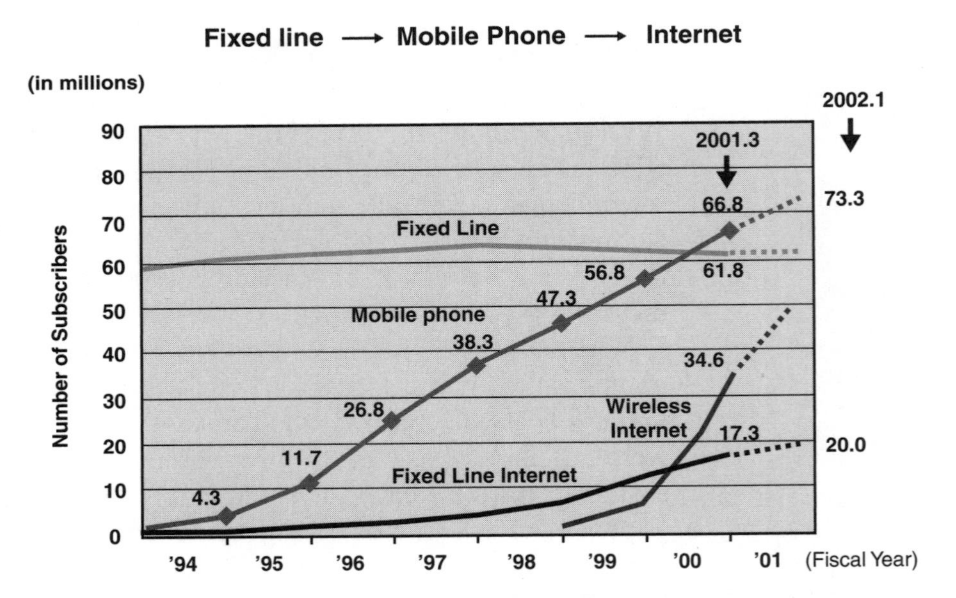

Although traditionally mobile business had been led by voice services, DoCoMo is looking to expand even further into nonvoice services. According to the company's estimates, nonvoice traffic is expected to account for approximately 50 percent of total traffic in 2005.

In terms of ubiquitous services, again there is significant growth potential. Anything that can move is a potential target for mobile communication services, which DoCoMo suggests should provide great opportunities for DoCoMo's networks to support person-to-machine and machine-to-machine communications. For example, in 2010 there will be an estimated 100 million cars in Japan, 60 million bicycles, and 50 million portable PCs, all of which could be potential users of mobile communications services.

DoCoMo is already involved in multimedia wireless with its benchmark i-mode service, which accounts for 18 percent of the company's revenues. i-mode was launched three years ago and at that time, DoCoMo's top management was skeptical about its value. Nevertheless, they invested in its development and have been rewarded with a large slice of the Japanese market. DoCoMo now has 31 million i-mode subscribers in Japan, which is equivalent to a 60 percent market share. The fastest growth is illustrated in Figure 1-3. Subscribers pay for access to the service and for selected services from content providers.

It is useful to note how i-mode has evolved from the early days, enhanced by the introduction of Java technology, which enabled secure transactions and the move into location-related information services. *Freedom of Mobile-Multimedia Access* (FOMA), NTT DoCoMo's 3G service, launched in October 2001, is the latest development in that evolution, and offers extensive mobile multimedia services. He outlined some of the potential applications of FOMA, including a video clipping service, and the location-related i-area service, enabling users to access local information such as weather forecasts or restaurant availability.

The company launched its personal navigation service using GPS in January 2000, with applications including the provision of information on the local traffic situation for car drivers. DoCoMo also launched a music distribution service in January 2001, enabling subscribers to access an extensive music database and retrieve and store music locally.

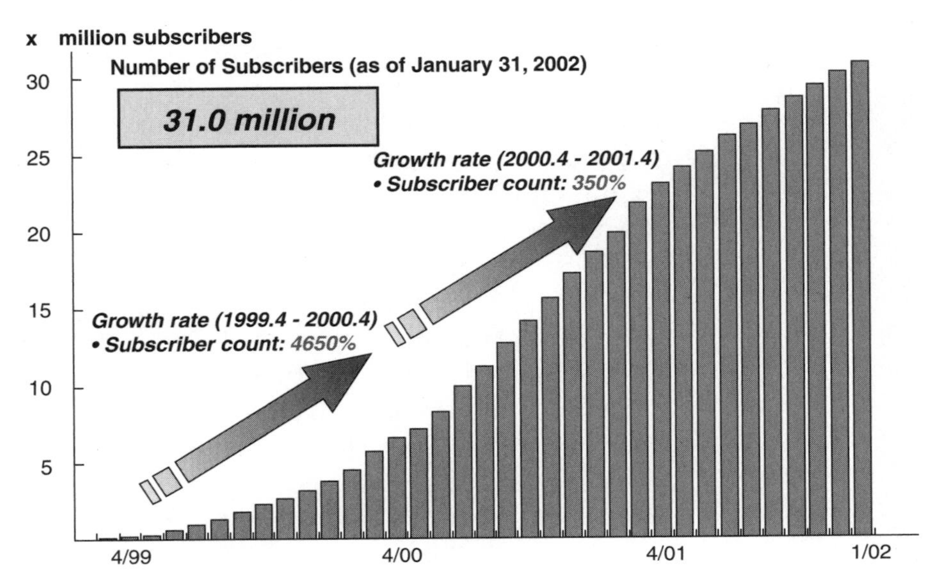

Figure 1-3 Number of sites registered with search engine OH! NEW (Source: NTT DoCoMo

An advertisement distribution service was launched almost two years ago, enabling advertisers to transmit timely information to users at a low cost. Subscribers can receive information tailored to the personal data given at the time of registration. Other applications include environmental monitoring and *point of sale* (POS) systems for vending machines.

The Scene Is Set

With an understanding of the relationships between the key players in the wireless sector, the potential of wireless messaging, the underlying business drivers behind it, and a clear idea of how the recent experiences of a number of players in the global market are shaping the perceptions of the wireless industry, we can now shift topics and look at the future of wireless messaging in the next chapter.

CHAPTER 2

Technology and Market Overview

Expect stunning growth in mobile communications. The number of mobile connections worldwide will increase from almost 727 million at the end of 2001 to more than 1.7 billion by the beginning of 2005 according to Goldman Sachs. The world will exceed the billion-user mark for mobile phones worldwide by 2003. By 2005, it is predicted that 500 million mobile devices will offer Internet access, a figure that will outstrip *personal computers* (PCs).

Of course, these are only numbers, but manufacturers and investors are betting that they will be more right than wrong. But will the user come if they build it? Will the data adoption take place because next-generation networks are in place? Will the data transmission speeds promised be achieved and give users the experience necessary for them to use wireless data regularly and be willing to pay for it? All of these questions still need to be answered.

For a capital-intensive industry, the wireless network industry is surprisingly unconcentrated. Most markets have between 5 and 10 competitors, and as of the end of 2000, there were approximately 937 carriers in the world, which was an increase of 175 percent from the 341 at the end of 1994. For mobile application developers and providers, a key element is associated with this statistic. Network complexity is unlikely to disappear in the near term, so applications will generally need to utilize extensive middleware or be standards based in order to offer the end user seamless access. For infrastructure application providers, this indicates that business models should be geared toward the volume of usage rather than the quantity of networks built in order to maximize value. One reason why carriers are moving to include mobile data capabilities is that voice *average revenue per user* (ARPU) has fallen at approximately 10 percent a year during the past decade. This trend is likely to continue.

It is important to present and explain the various technologies discussed in this book. During the last 12 months, sentiment toward both wireless carriers and equipment vendors has undergone substantial swings, as enthusiasm over the prospects for the transition from today's voice-centric *second-generation* (2G) wireless networks to a new generation of data-enabled, *2.5-generation* (2.5G) and *third-generation* (3G) wireless networks has ebbed and flowed.

The wireless community has assaulted users with a plethora of complex numbers and acronyms in its discussion of the technologies under consideration for wireless network upgrades. To simplify the discussion, we refer to the current generation of digital wireless networks in use today as 2G networks. Upgrades to these existing networks using the same core systems and the same wireless frequencies have often been referred to as 2.5G network upgrades. Meanwhile, the construction of entirely new greenfield networks using new spectrum often acquired in government auctions has generally been referred to as the development of 3G networks.

3G has also been used to describe a broad range of higher-speed data networks whether these networks are upgrades to existing systems or new greenfield deployments using new spectrum. The *International Telecommunications Union* (ITU) requires standards to support 2 Mbps data rates in a stationary environment and 384 Kbps in a mobile environment in order to merit the designation of a 3G standard. These standards include *Wideband Code Division Multiple Access* (W-CDMA), the CDMA2000 1x family, and a variant designed for the Chinese market termed *Time Division Synchronous CDMA* (TD-SCDMA). We have broadened the definition of 3G to refer to data upgrades that are capable of consistently delivering higher speeds of more than several hundred kilobits of throughput per second, regardless of whether they are leveraging the legacy network in the existing frequencies or brand new systems.

Wireless service operators are in the process of implementing, or planning to implement, upgrades to their current 2G networks. Several technology choices are available. The choice of 3G is expected to introduce significant competitive differentiation among wireless service operators in their ability to timely and effectively offer data services. Although the new 3G networks will provide enhanced data speeds, making the user experience much more enjoyable, the primary motive for 3G implementation is the additional voice capacity the technology provides.

The wireless sector has suffered from a number of high-profile delays in next-generation network infrastructure deployment timelines. The overzealous initial expectations are now adjusting to

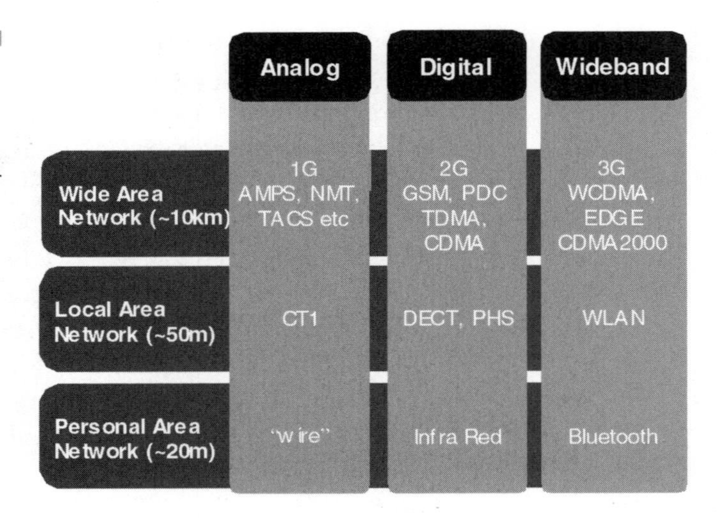

Figure 2-1 Overview diagram and technologies (Source: Motorola)

a more sanguine view of timelines for the rollout of data-enabled networks. The timing of some of the upgrades remains uncertain, however, since network complexity, difficulty establishing service levels, handset constraints, and high-cost challenges remain.

This chapter discusses *local area network* (LAN) technologies known as *wireless LAN* (WLAN), Bluetooth and *wireless wide area network* (WWAN). Both WLAN and Bluetooth technologies are discussed here because they are commercially available today, although they cannot be used in the same technology as a cellular or WWAN (the network that a cellular handset operates on today). In the not too distant future when handsets will have internal WLAN and Bluetooth capabilities, they will be able to utilize each technology for communication. Thus, this chapter compares the technologies alongside the cellular networks. Figure 2-1 provides a brief overview of these technologies and their characteristics.

Figure 2-2 provides an overview of how the different technologies complement each other.

Figure 2-2
Coverage versus data rate (Source: Ericsson)

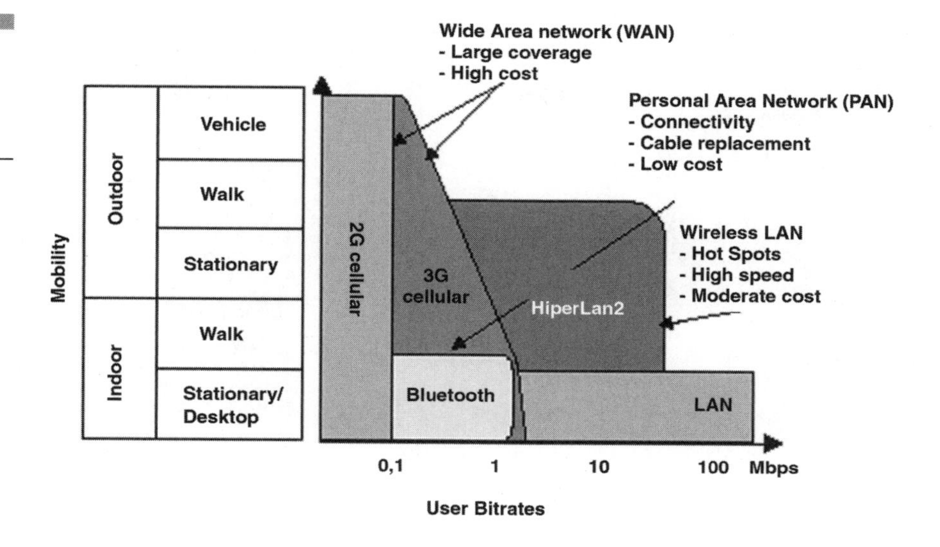

How Will These Networks Affect Messaging?

Technology has a significant impact on the fundamentals of the wireless business. In particular, the digital air interface that is chosen to deploy wireless service affects the *quality of service* (QoS) and *return on investment* (ROI) for the wireless service operators. In general, strategic and well-timed technology decisions can help wireless service operators:

- Lower operating costs by increasing network capacity and making more efficient use of spectrum.
- Increase ARPUs by enabling operators to provide new types of wireless data services such as *Short Message Service* (SMS) while enhancing that service to *Enhanced Messaging Service* (EMS) and *Multimedia Messaging Service* (MMS).
- Reduce churn by improving QoS.

Figure 2-3 illustrates what new technology will support, accompanied with format characteristics.

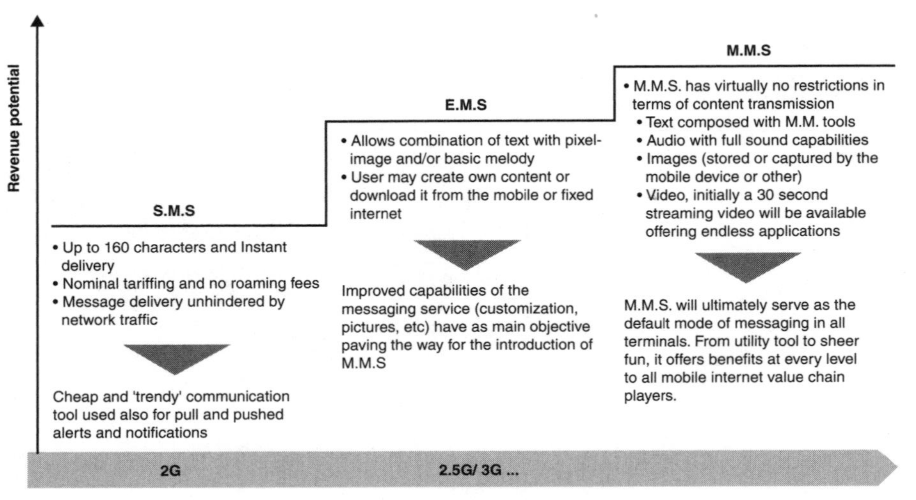

Figure 2-3 Messaging formats and characteristics with associated technology (Source: Xfera)

This chapter focuses on the traditional cellular/*personal communications services* (PCS) wireless service operators and does not refer to specialized wireless data networks (such as Motient, Cingular Interactive, ReFlex [paging], Metricom, and so on). The wireless industry went through its first major technology upheaval in the 1990s when operators chose to digitally upgrade the path from analog cellular to digital. Rather than mandating a single standard akin to the European Union's *Global System for Mobile Communications* (GSM), the United States let competition decide, for better or for worse. Whether or not the wireless service operators knew it then, their decisions to go to *Time Division Multiple Access* (TDMA), GSM, or CDMA have had an impact on their businesses since and will have a much more material impact on their businesses as they convert to 3G. The choice of digital technology to date has affected spectrum allocation due to the different channelization and frequency reuse of the various technologies in addition to a host of other issues that will not be deeply discussed here (such as time slots per channel, guardbands, Erlangs per site, and so on).

WWAN Technologies: Including CDMA2000 1xRTT, GSM, GPRS, EDGE, and W-CDMA

First Generation (1G)—Analog

In 1978, *Advanced Mobile Phone Service* (AMPS) started operating in North America. That year, AT&T labs in Newark, New Jersey, and Chicago, Illinois, rolled out the first analog-based cellular telephone services. The Chicago system operated in the newly allocated 800 MHz band and was made up of 10 cells covering approximately 21,000 square miles. This early network, which used large-scale integrated circuits throughout, a dedicated computer controller and switching system, and custom-designed mobile telephones and antennas, proved that cellular could work.

AMPS channels have a bandwidth of 30 KHz, which translates into a channel allocation per carrier of 832 (25 MHz/30 KHz = approximately 832 channels). Of these 832 channels, 406 are reserved for transmit, 406 are reserved for receive, and 20 are reserved for control channels; thus, if reuse was not an issue, 406 simultaneous conversations could take place on each carrier's bandwidth.

Overall, analog systems provide the least effective use of frequency spectrum, and analog handsets consume the greatest amount of power. This is why cellular carriers are no longer deploying new analog systems and the number of worldwide analog subscribers continues to decrease. In 2001, there were an estimated 40 million analog subscribers. By 2006, that number is forecasted to be down to less than 24 million. Most analog subscribers reside in smaller and rural communities and have little reason to switch to digital.

2G—Digital

Digital cellular systems have many advantages over their analog cellular counterparts, but the most important of these is increased capacity. In the 1980s and early 1990s, as cellular use quickly rose, operators realized that if they didn't act, the capacity of their analog cellular systems would soon be exceeded in some markets. Digital cellular was a way for operators to not only increase capacity, but also to add new revenue-producing features, such as data, caller ID, SMS messaging, and a host of others.

Digital cellular systems fall into two general classes: TDMA and CDMA. TDMA systems work by splitting a cellular channel into several time slices and putting multiple voice conversations on these time slices. TDMA technologies include TDMA, GSM, *Personal Digital Cellular* (PDC), and *Integrated Digital Enhanced Network* (iDEN). CDMA systems enable many voice conversations to share one channel, but these conversations are separated by assigning a unique identifying code to each one. CDMA-type systems include CDMA, CDMA2000, and W-CDMA.

TDMA, GSM, D-AMPS, and PDC GSM development dates back to 1982. It was started by a group of 26 European national phone companies. The *Conference of European Postal and Telecommunications Administrations* (CEPT) sought to build a uniform cellular system around 900 MHz across Europe. In 1989, The *European Telecommunication Standards Institute* (ETSI) took responsibility for further developing GSM. In 1990, the United Kingdom asked for and was granted the right to use GSM at higher frequencies, and a *Digital Cellular System* (DCS) service called DCS1800 was born. As the name implies, DCS1800 operates at 1,800 MHz. GSM is divided into 8 time slots per 4.62 ms frame, providing the capacity for 8 simultaneous full-rate traffic, or voice, channels (half-rate would result in 16 channels).

Japan and North America watched as GSM spread across Europe and realized that they also needed a digital system. Instead of adopting GSM, these two countries developed their own systems. North America chose *Digital AMPS* (D-AMPS), and Japan chose PDC.

D-AMPS, or IS-54, as it is often referred to, uses a technique of converting a call to digital and then intermixes the call with two other calls. Three digital calls then occupy the space that had previously been occupied by one analog call. This technique, called TDMA, greatly increased capacity and enabled a cellular carrier to convert any of its systems' analog voice channels to digital as needed.

In 1994, the IS-136 TDMA standard was introduced. IS-136 replaced IS-54 while still maintaining backward compatibility with the older standard. IS-136 includes a digital control channel that allows the use of new features on 800 MHz TDMA band that were similar to the ones starting to be promoted on the PCS 1,900 MHz band.

In Japan, another system based on TDMA was adopted. The PDC system is very similar to D-AMPS. Like D-AMPS, the Japanese system overlaid the existing analog system already in place so that compatibility with existing subscribers could be maintained. Like the American system, three digital time slots were placed within each analog channel, providing a threefold increase in system capacity. PDC is offered in two frequency bands: 810 to 826 MHz receive/940 to 956 MHz transmit and 1,429 to 1,453 MHz receive/1,477 to 1,501 MHz transmit.

D-AMPS and PDC are divided into six time slots per 40 ms frame, providing the capacity for three simultaneous full-rate voice channels as each full-rate channel occupies two slots. For D-AMPS, half-rate channels were later defined, which only occupied one slot per frame, allowing six simultaneous conversations. These digital techniques significantly increased traffic capacity over the analog systems.

CDMA In 1994, Qualcomm proposed a spread spectrum scheme to increase system capacity. CDMA, which was built on an earlier proposal, would be all digital and promised 10 to 20 times the capacity of existing AMPS cellular techniques. CDMA2000 1x, which is currently available, aims to offer CDMA carriers nearly double the voice capacity of their current CDMA-One, or 2G CDMA, networks. CDMA-One networks offer up to 10 times the voice capacity of analog networks, according to the CDMA Development Group. CDMA2000 1x is up to 20 times the voice capacity of analog networks.

In CDMA, each user is assigned a *pseudonoise* (PN) code to modulate transmitted data. The PN code is a long sequence of ones and zeros (similar to the output of a random number generator of a computer). The numbers are not really random; they are generated using a specific algorithm. Because the codes are nearly random, there is very little correlation between the different codes and between a specific code and any time shift of that same code. Thus, the distinct codes can be transmitted at the same time and at the same frequencies, and the signals can be decoded at the receiver by correlating the received signal (which is the sum of all transmitted signals) with each PN code.

CDMA systems are used in several parts of the world, but the majority of CDMA subscribers are located in the United States, Brazil, Japan, Korea, and China. Because Qualcomm owns the patents on many aspects of CDMA technology, it generally receives royalties for all the CDMA handsets and base stations produced, whether or not they actually use Qualcomm chips. At the conclusion of 2001, almost 12 percent of worldwide cellular subscribers used CDMA, whereas in the United States, CDMA subscribers made up approximately 36 percent of the total subscriber base. Data rates are up to 14.4 Kbps and average 9 to 14 Kbps.

Personal Handy Phone System (PHS) Soon after PDC was developed, Japan developed another cellular system known as the *Personal Handy Phone System* (PHS). PHS is considered to be a low-tier TDMA technology. It is classified by some as a *Digital Cordless Telephone* standard, which competes with *Digital European Cordless Telephone* (DECT) and *Cordless Telephone 2* standards (CT2/CT2+). In Japan, this system provides access to business phone systems, residential base stations, and the public network, but the system only operates for subscribers moving up to pedestrian speeds. As a result, PHS is positioned between cellular and cordless systems.

The PHS radio interface is based on a TDMA/*Frequency Division Multiple Access* (FDMA) with time-division duplex. It operates with 77 radio slots in the 918 MHz band, which use a 300 KHz wide carrier and support 4 channels for 308 total channels ($4 \times 77 = 308$). Unlike typical AMPS, TDMA, GSM, or CDMA cellular/PCS mobile, PHS infrastructure uses many small base stations of either three traffic/one control channel or seven traffic/one control channel configurations.

iDEN This private mobile radio system comes from Motorola's *Land Mobile Products Sector* (LMPS) and was launched in 1994. iDEN technology is currently available in the 800 MHz, 900 MHz, and 1.5 GHz bands. iDEN utilizes a variety of advanced technologies, including state-of-the-art vocoders, M16QAM modulation, and TDMA. It enables *Commercial Mobile Radio Service* (CMRS) operators to maximize the dispatch capacity and provides the flexibility to add optional services such as full-duplex telephone interconnect, alphanumeric paging, and data/fax communication services. iDEN systems operate in the United States, Canada, Argentina, Israel, Japan, Singapore, China, the Philippines, and Colombia.

PCS By the mid-1990s, systems in the United States were again approaching full capacity, especially in the more densely populated cities. After much study, the *Federal Communications Commission* (FCC) decided that it would open the 1,850 to 1,990 MHz band to a new type of cellular service called PCS. Spectrum in the newly designated band was auctioned from December 1994 to January 1997. Unlike the 800 MHz band, which allowed space for two competing carriers, the PCS band could contain up to six carriers, which were designated as blocks A through F. These blocks, unlike those in the 800 MHz cellular band, were offered in two different sizes, enabling both large and small carriers to effectively compete.

The A, B, and C blocks have a bandwidth of 15 MHz and allow for 300 channels each. The D, E, and F bands have a bandwidth of 5 MHz and have space for 100 channels each. When combined, PCS has a total of 1,200 channels, compared with approximately 830 channels for 800 MHz cellular.

In addition to channel capacity, PCS has another capacity benefit over 800 MHz cellular. Because PCS operates at a higher frequency than cellular, the range from the base station to handset is generally reduced and cell coverage is about 40 percent smaller. This allows for an increased density of PCS cell sites for a given area, which, in turn, allows for a greater system capacity. Overall, this gives PCS a two to three times greater capacity advantage when compared with 800 MHz cellular; unfortunately, this also means that more base stations are required for deployment and results in higher capital expenditures.

The Next Generation: The Upgrade

The impact of wireless service operators' spending choice toward 3G will manifest in spectrum, infrastructure, handset costs, and competitive differentiation. Clearly, technology matters.

Network capacity increases only through additional spectrum or technological improvements in the radio interface, which allow more bits to be transmitted over the same amount of spectrum. Given the increase for voice and data traffic, it is clear that operators need to add capacity as time goes by. The following are the main drivers for this traffic growth:

- **Voice** The number of wireless subscribers continues to increase approximately 20 percent a year. In addition, the average user is consuming more voice minutes per month. The total number of voice minutes used by a U.S. wireless service provider is anticipated to increase approximately 33 percent over the next four years. Before wireless moves to *voice over IP* (VoIP), circuit-switched voice telephony will continue to consume more spectral resources than packet data.
- **Data** The amount of data traffic should grow quickly from the use of applications and services focused on communications, information, entertainment, and commerce. Messaging services, such as SMS, IM, e-mail, EMS, or MMS, are anticipated to consume the majority of wireless data traffic in the future primarily because of the *person-to-person* (P2P) (communication) experience.

Wireless services operators must upgrade their network for two main reasons:

- **Voice capacity** Wireless operators that upgrade their networks from analog and 2G digital technologies will gain the increased capacity that 2.5G and 3G networks offer. Although most wireless operators state that they are not currently experiencing spectrum shortages (even though users throughout the world in major urban areas consistently report drop calls during peak conditions), the need to make better use of already-owned spectrum will increase as operators fully load their networks with voice traffic through the addition of new subscribers. Furthermore, given the shortage of new spectrum for wireless services and the cost of obtaining spectrum at auctions, wireless service operators are compelled to implement technology solutions that will make more efficient use of spectrum. Seemingly simple solutions such as shutting down spectrally inefficient analog networks are currently prohibited by the FCC in the United States, although this set-aside rule may disappear. In short, the operators have a clear economic incentive to upgrade.

- **Data capacity** As network operators deploy 2.5G networks and their higher data transfer rates, data usage will increase through new applications that these speeds allow, creating an enhanced user experience that will be combined with more sophisticated color phones/handsets. New applications will also likely require additional capacity because of the increased amount of traffic carried on the networks, such as MMS. To remain competitively differentiated and avoid missing out on new data revenue streams, wireless service operators will need to upgrade their networks in order to be in a position to offer such services. Similar to the historical example of PCs and the Internet, more data-intensive applications will probably not emerge until networks and technology that are capable of supporting such applications exist.

Data Services May Help Boost ARPU Wireless service operators are looking for ways to increase their ARPU. Wireline operators have seen tremendous pricing pressure on pure voice services over the past decade. Ultimately, wireless voice pricing is likely to mirror the slow, steady decline seen by the wireline operators over the past several years.

Wireless service operators are looking to provide additional services to their subscriber base beyond pure voice service. In order to provide these value-added, ARPU-raising services, such as messaging and Internet access, wireless service operators must improve the data rates that their networks can offer the handsets and other devices. Current 2G networks require a long connection time—much like a dial-up modem at home—and offer meager data rates of approximately 14 Kbps. New 2.5G (GPRS and IS-95B) and 3G (CDMA2000 and W-CDMA) network technologies will offer always-on connectivity and high-speed data rates.

New enhanced data rates equal new wireless data services and thus new revenue streams. Currently, SMS revenues from services

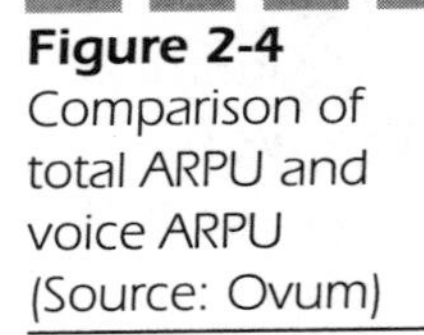

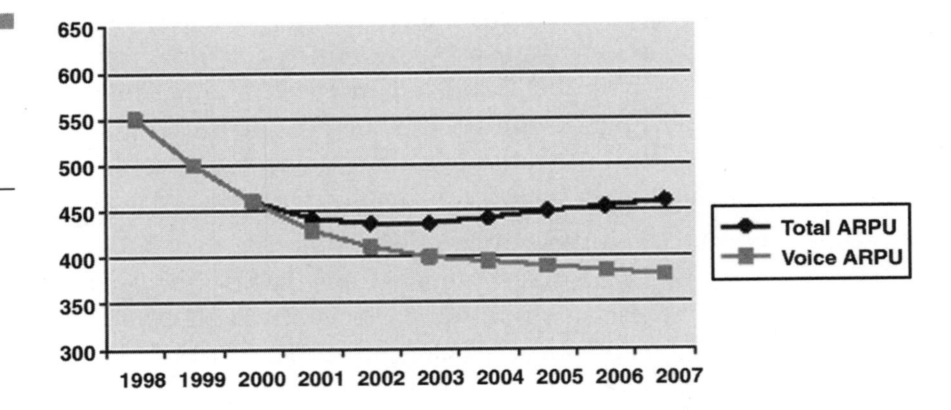

Figure 2-4 Comparison of total ARPU and voice ARPU (Source: Ovum)

comprise approximately 15 percent of sales in the majority of wireless service operators in Europe. The next generation of the WAP standard, which will add support for *Hypertext Markup Language* (HTML) documents and the ability to send MMS messages, is likely to facilitate the adoption of data applications.

Figure 2-4 illustrates how industry observers believe the ARPU tail will play out. As you can see, wireless data for the wireless service operators is quite important and is seen as a substantial component of overall revenue. Using the Globe in the Philippines as an example, 35 percent of their total revenue is from SMS. That is significantly above any other wireless service operator. Once MMS is capable via the next-generation networks, pictures acting as the major catalysts should help other wireless service operators increase their percentage of revenue from data.

Understanding Packet- and Circuit-Switched Network Technology There are two ways to make a wireless connection from your handset to the data a user may want to obtain:

- Circuit-switched technology
- Packet-switched technology

Circuit technology uses a dedicated network approach, whereas packet technology implies the concept of a shared network. In the simplest terms, a circuit-switched connection is like dialing up your

Internet service provider (ISP) from home over traditional copper phone lines, whereas a packet-switched connection is like having a cable modem or *Digital Subscriber Line* (DSL) connection.

Circuit-switched connections are very bandwidth inefficient for the network operator because an entire voice channel is dedicated to the data connection whether or not the user is transmitting data, thus these networks are not optimal for wireless data.

On the other hand, packet-switched connections enable data to be broken into pieces and routed along unused network channels, which is ideal for messaging applications. If a circuit-switched connection is lost, the entire process must be undertaken again; if a packet-switched connection is lost, only the data that did not get through is resent, and the entire message is reassembled at the end.

Advantages of Being Always On It is also important to understand that the circuit-switched network tends to be slower because it requires extra dial-up and connection times. In fact, the circuit-switched networks require devices to dial a number, connect, and log on, whereas packet-switched networks are always on and provide virtually instantaneous access. For the wireless service provider, packet networks offer more network efficiency relative to circuit-switched data networks.

Migration Path We have identified the following network migration paths to 3G for the wireless service providers, depending on their current technology:

- **For CDMA operators** The path is from CDMA-One (IS-95) to CDMA2000 1xRTT and then to *CDMA2000 1x evolution* (1xEV) (both *data only* [DO] and *data and voice* [DV]).
- **For GSM operators** The path is from *General Packet Radio Service* (GPRS), which is perhaps followed by *Enhanced Data Rates for Global Evolution* (EDGE), and then leads to W-CDMA.
- **For TDMA operators** A few paths are possible. The first, highlighted by AT&T Wireless, involves constructing a GSM/GPRS overlay on its network, implementing EDGE, and then upgrading to W-CDMA. Another path involves upgrading to EDGE and then upgrading to W-CDMA. The majority TDMA

operators will choose the first path (migrating to GSM/GRPS first).

- **TDMA-GSM convergence** In the meantime, a new 2G path for TDMA operators has emerged: *GSM American National Standards Institute (ANSI) Interoperability Team* (GAIT). GAIT is a new network standard that marries TDMA and GSM to enhance roaming, but preserves TDMA network technologies. In some markets, GAIT may eliminate the need for a GSM/GPRS overlay, but it will not obviate the need to choose a path to W-CDMA. GAIT is a software upgrade that requires a single dual-band phone; no new base stations or *Mobile Switching Centers* (MSCs) are needed.

Figure 2-5 illustrates the network upgrade paths for all cellular technologies.

There has always been a heated debate as to which technology is 3G and which is actually next generation. For the purposes of this book,

- 2.5G is defined as those technologies offering data rates higher than 14.4 Kbps and less than 120 Kbps.
- To many, these intermediate speed cellular data services may just seem as a stepping stone to things to come. However, the

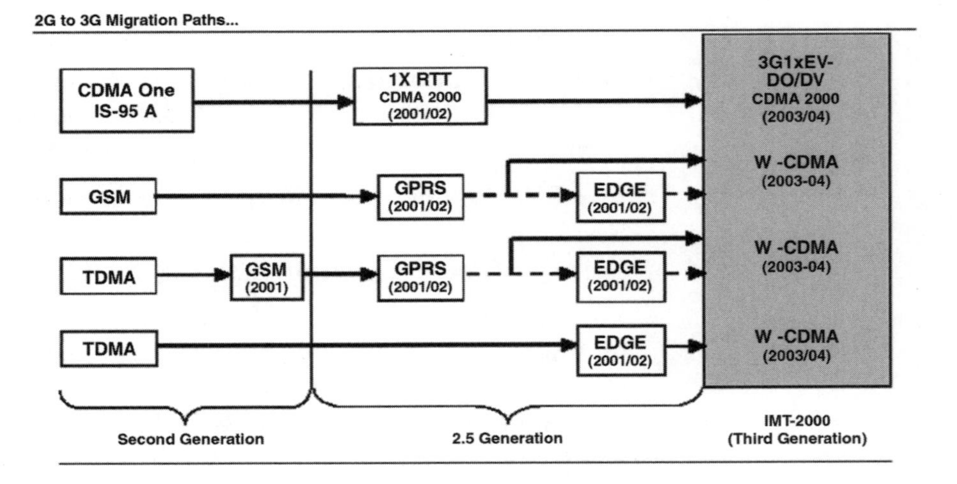

Figure 2-5 2G to 3G migration paths (Source: Motorola)

reality will likely be that these data services will be with us much longer than some mobile operators want to believe.

2.5G technologies fit into cellular nicely for a variety of reasons. First, 2.5G technologies are packet based as opposed to 2G data services, which were generally connection based. This allows for always-on services. This is similar to what most users are used to with the wired Internet. In addition, since no real connection needs to be established, latency in sending data is greatly reduced.

Second, since 2.5G services don't require new spectrum or new 3G licenses, the carriers' cost to deploy these services is modest; therefore, the costs to the consumer will be modest as well. High cost for service is one of the biggest obstacles 3G faces when deployed.

Third, enabling handsets for 2.5G services is fairly inexpensive. Typically, from a silicon standpoint, adding 2.5G functionality adds less than $10 to the handset cost.

As of the writing of this book, most GSM service providers have started their deployment of GPRS. Although results vary per service provider, most GPRS users are receiving data rates of 25 to 35 Kbps to the handset and rates that are less than half that from the handset.

In the CDMA side of the house, CDMA2000 1x uptake has been occurring quite strongly in North Korea. Qualcomm claims that there are currently over 12 million 1x subscribers in Korea. In the United States, Sprint, Alltel, and Verizon are all currently deploying 1x networks. These carriers hope to complete these deployments by the end of this year.

GPRS GPRS is a packet-based protocol that supports data rates over 100 Kbps using eight time slots, but other rates can also be supported using fewer time slots. GPRS can dynamically change its resource time slot allocation to give priority to voice traffic. It also has the capacity to allocate SMS traffic. Should the network signal channel be running at peak, GPRS can dynamically manage channel allocation and send some or all of the SMS traffic over GPRS channels alternatively.

The migration to GPRS does not expand voice capacity. In fact, in the short term, GPRS upgrades may increase network congestion, as data traffic will occupy channels that could be used for voice traffic.

This could force operators to expand capacity by adding base stations as voice and data traffic increases. Note that leading equipment vendors such as Nokia and emerging startups such as mDiversity are actively developing means to increase the voice capacity of 2G GSM networks without requiring installations of new base stations.

GPRS can operate in one of several classes. The class determines how many time slots are allocated to downlink data (to the handset) and how many are allocated to uplink data (from the handset). Of the eight time slots in a normal GSM frame, a normal voice call uses three slots—one for the forward link, one for the reverse link, and one for the monitor slot (a slot for the handset to monitor neighboring cells).

Because of power and radiation concerns, most initial deployments of GPRS will be for class 2, 3, or 8. As GPRS becomes more popular, customers will demand higher data rates, and these higher classes will gain greater popularity. GPRS data rates have been operating between 25 to 40 Kbps.

An initial GPRS upgrade is relatively inexpensive because the migration pattern involves primarily software upgrades. Little new hardware is required. New software is necessary in the *Base Station Transceiver* (BST), the *Base Station Controller* (BSC), and in the network databases such as the *Home Location Register* (HLR) and *Visitor Location Register* (VLR). Indeed, one of the larger costs involved in migrating a network to GPRS is the cost of the required software licenses.

GPRS occupies the voice capacity to transmit data, but it does not increase voice capacity. The successful introduction of data traffic could lead operators to invest in more infrastructure capacity. Thus, the rollout of GPRS could be a significant growth driver over time for leading GSM equipment vendors such as Ericsson and Nokia. These vendors are also actively looking to increase the voice capacity of GSM networks without adding base stations. The major vendors are developing enhancements internally, such as Nokia's *Adaptive Multirate* (AMR) coding, and evaluating external methods, such as mDiversity's technique for reducing shadow fade.

GPRS networks require a small amount of additional hardware in order to transport data transmissions. A *packet control unit* (PCU) at the BSC splits the packet data from circuit calls and sends it to the GPRS support nodes. The *Serving GPRS Support Node* (SGSN) essentially mimics the functions of the MSC for packet networks in that it detects and tracks the position of new data users in its geographic region, queries subscribers' HLRs, and delivers packets to the proper BSC. The *Gateway GPRS Support Node* (GGSN) is an interface to external *Internet Protocol* (IP) networks and functions in many ways like a router.

Pros

- A packet data network is only active when data actually needs to be sent or received, which keeps costs lower for the user.
- There is dynamic allocation of time slot resources.
- It reduces load on the signaling channel by handling some of the SMS traffic.
- IP and X.25 are fully supported; all existing Internet tools are supported.
- Always-on virtual connections are supported.

Cons

- Initial GPRS phones will only support lower-speed versions of GPRS using four or less total time slots.
- The actual transmission speeds observed may be well under theoretical maximums.
- GPRS cannot match the spectrum efficiency of EDGE or CDMA2000.
- Power consumption and radiation will increase as the number of transmit time slots increase.

EDGE EDGE is a high-speed mobile data standard and is considered an evolution of GPRS. EDGE achieves packet data transmission speeds of up to 384 Kbps using eight time slots with a data rate of up to 48 Kbps per time slot. Higher speeds may be possible under ideal conditions, but the radio channel is shared so throughputs could be much lower as well.

EDGE gives incumbent GSM operators the opportunity to offer data services at speeds near those available on *Universal Mobile Telecommunications System* (UMTS) networks while still using the current 200 KHz GSM frequencies. EDGE achieves its faster data rate by employing an advanced modulation technique (8 *phase-shift keying* [PSK]) similar to what UMTS uses and is therefore considered an evolutionary step toward UMTS.

For handset manufacturers, the migration path from GPRS to EDGE will not be simple. Handsets will need to have extra processing power to handle EDGE modulation and added memory for use in high-speed data buffering. It is expected that the first handsets to support EDGE may incorporate it in the reverse link (to the handset) only. High data rates will be available from the base station to the phone, but not in the other direction. As prices for EDGE decrease, phones will start to incorporate the technology in both directions.

To date, support for EDGE has mostly evaporated throughout Europe. Instead, most European companies want to invest their limited resources in W-CDMA. In the United States, both AT&T Wireless and Cingular Wireless have committed to EDGE. Deployments are slated to start as early as 2003. A few chipmakers, such as PrairieComm, have committed to making EDGE chipsets for handsets, and a few handset makers have committed to making EDGE handsets as well. Still, it's expected that EDGE handsets will be somewhat more expensive than GPRS handsets due to their higher processing demands and limited supply.

Pros

- Its spectrum efficiency is more than double that of GPRS.
- Data rates approach 384 Kbps; it is fast enough to transfer streaming audio and video.
- The use of EDGE could eliminate most of the need to deploy costly W-CDMA.

Cons

- Initial handsets will be more expensive than GPRS, and service charges are also likely to be higher.

- It sometimes requires new infrastructure hardware, or at a minimum, a software upgrade and new radios.
- Handset power consumption may be high, as might be handset radiation emissions.

CDMA2000 1xRTT CDMA2000 1xRTT is an evolutionary step from IS-95A and uses the same 1.25 MHz of bandwidth as the earlier technology; therefore, it can be deployed on existing spectrum. 1x improves upon IS-95A and has more sophisticated power control, new modulation on the reverse channels, and improved data-encoding methods. For a relatively modest investment in infrastructure, CDMA carriers can offer IP-based packet data rates of up to 144 Kbps with 1x. They typically offer 50 to 70 Kbps, which is nearly double the voice capacity of existing spectrum.

1x has been rolled out successfully in South Korea. All American CDMA carriers are hoping to deploy 1x nationwide by the end of 2002. Adding 1x capabilities to a CDMA-One handset adds very little to the cost, and any increases are almost exclusively confined to the baseband section and some added memory. Despite this, initial CDMA2000 handsets have carried a $70 to $100 premium in the South Korean market, but this is rapidly decreasing as these handsets become more popular and their chipsets become more widely available. During 2002, most CDMA service providers planned to totally phase out their use of CDMA-One handsets and sell CDMA2000 1x handsets exclusively.

The migration to CDMA2000 1x requires upgrades in software, hardware, and the core network. The most expensive components are the channel cards and upgrades of the core network to handle increased data throughput. Software makes up the remainder of the costs. Analysts estimate that the initial investment needed to enable a current CDMA network to have CDMA2000 1x data capabilities is approximately $2 to $4 per user. As previously noted, this investment increases both data throughput and voice capacity and thus may require less incremental investment over time by the operator than the GPRS path, which does not increase voice capacity.

Pros

- Modest investment is required for upgrading existing IS-95A infrastructure.
- It can coexist with IS-95A and IS-95B networks.
- It uses existing spectrum to offer packet data services, provides for an increase in capacity, or does both.
- It provides a very cost-effective migration path for carriers.
- There is a minimal increase in handset chipset cost over IS-95A and IS-95B.
- Generally, it has higher data rates than GPRS solutions, but has probably less than EDGE.

Cons

- Coverage of CDMA2000 will generally be limited to Korea, China, India, Japan, and the United States.
- CDMA2000 chipsets and handsets are generally more costly than GPRS, but they may be less costly than EDGE.

3G

3G refers to the new wave in the evolution of the wireless phone and generally refers to technology that supports data rates in excess of 144 Kbps. 3G is not made up of just one technology or standard; it contains many. Demands that are driving this new technology include the following:

- Global roaming
- Lower subscriber costs
- Increased system capacities
- Wireline-grade voice quality
- Multimedia services
- High-speed data
- Internet-like IP-based packet data services

The ITU is the telecommunications governing body overseeing worldwide communications. Back in the early 1990s, this group saw the need to develop a new set of standards that would provide the necessary vision to drive the next-generation communications products into the next millennium and beyond. This new standard is called *International Mobile Telecommunications-2000* (IMT-2000) and is the guiding force behind 3G.

As might be expected, it was not an easy task to produce a single worldwide 3G standard that all parties could agree on. This is why the ITU chose to adopt five air interfaces for its IMT-2000 standard. Each of these standards is an evolution of a 2G standard that is currently being used. Although the end result is not the single worldwide standard that the ITU might have hoped for, the standards chosen do simplify the migration paths for the incumbents and offer economic and geographic variety to the new 3G license holders.

Although the premise for 3G was relatively simple, it has turned out to be much more complex. High costs, equipment problems, macroeconomic slowdowns, and a general lack of demand have all slowed the spread of 3G systems worldwide. In Japan, where NTT DoCoMo rolled out its W-CDMA system last October, subscriber growth has been modest. NTT reported approximately 112,000 subscribers as of May 2002. Currently, only two handsets are available for use on its network: the *Freedom of Multimedia Access* (FOMA) N2001 made by NEC and the FOMA P2101V manufactured by Matsushita.

W-CDMA W-CDMA is an evolutionary step up from GPRS and EDGE. It uses a wide 5 MHz carrier bandwidth, 4.097 Mcps chip speed, and CDMA air interface to attain high-speed DV services. Future versions may use a 10, 15, or 20 MHz bandwidth, but with spectrum in short supply, systems such as this are not very likely. W-CDMA is not interoperable with 2G+ technologies and will only operate on new IMT-2000 frequencies; however, there is some talk of running W-CDMA on the PCS frequencies.

Current W-CDMA handsets are single-band, single-mode W-CDMA only, but versions that also support GSM are expected shortly for use in Europe. NTT DoCoMo is resisting pressure to add

PDC i-mode support to its W-CDMA phones; instead, it is choosing to market its FOMA as a separate service to its current offerings. Data rates for W-CDMA are up to 144 Kbps.

Pros

- High stationary data rates in excess of 1 Mbps are possible, but real-world rates are much lower.
- A large subscriber base could potentially reduce costs, but the number of subscribers has so far been low.

Cons

- It requires new IMT-2000 frequencies, which will raise the cost for service.
- Dual-mode, dual-band handsets will be required for compatibility with non-3G service.
- Wide bandwidth reduces the number of channels available.
- Initial chipsets and the handsets made with these chips have been very expensive.

CDMA2000 1xEV-DO CDMA2000 1xEV-DO is a form of CDMA2000 that offers high-speed data connectivity with no voice support. However, the technology can be used in voice-centric devices, such as handsets, in combination with CDMA2000 1x. Although 1xEV-DO must operate in a separate *radio frequency* (RF) carrier than CDMA2000 1xRTT, it isn't a handicap, since most service providers will likely choose to keep voice and data separated, even with technology that enables them to coexist.

Currently, 1xEV-DO systems are being commercially deployed in South Korea and beta tested in the United States, achieving data rates of 800 Kbps to 1.4 Mbps. Current technology doesn't offer true seamless handoff between 1xEV-DO and 1xRTT systems, but this could be solved within the next few years. If the issues are resolved, it will be possible to receive a voice call with 1xRTT without losing a data connection and switch to a higher-speed DO connection when the voice call is complete.

Pros

- It offers data rates even higher than CDMA2000 1xEV-DV.
- It enables DV traffic to be segregated into different RF carriers.
- It enables DO services to be deployed at a relatively low cost.

Cons

- 1xEV-DO doesn't support voice. Two RF carriers are required to support both voice (1xRTT) and high-speed data (1xEV-DO).
- The standards for seamless connectivity to both a 1xRTT network and 1xEV-DO network simultaneously have yet to be worked out.
- Dual-mode 1xRTT/DO chipsets are likely to cost much more than single-mode 1xRTT chipsets.

CDMA2000 1xEV-DV CDMA2000 1xEV-DV offers potential data rates in excess of 2 Mbps at peak using an existing 1.25 MHz bandwidth. 1xEV-DV supports both high-speed data and voice.

From a technical perspective, the CDMA migration path may benefit from increased voice capacity and higher data throughputs before other solutions. As data takes off, these capacity and throughput benefits may lead to lower operating costs over time. In addition, the CDMA migration path may enable carriers to offer high-speed data services operating at an average of several hundred kilobits per second by 2004 or 2005. For existing CDMA carriers in particular, the upgrade to CDMA2000 1x and 1xEV may represent a relatively simple and elegant solution.

Pros

- It offers W-CDMA data rates in currently allocated spectrum.
- It has high spectrum efficiency.
- Both voice and data is supported in one RF carrier.

Cons

- 1xEV-DV is not yet a fully approved standard and won't likely be deployed until after 2005, if then.

- If 1xEV-DO can be successfully combined with 1xRTT, 1xEV-DV is no longer very important.

Table 2-1 highlights the different network technologies, their migration paths, and their advantages and disadvantages.

Technology Briefs

Clearly, GPRS has gained considerable commercial momentum around the globe and has significant potential for improving economies of scale over time. GSM vendors suggest that if data usage does ramp and additional capacity is needed for GPRS, carriers will be eager to make those additional investments to either EDGE or UMTS. After experiencing some significant initial delays in deployments, GSM carriers across Europe and Asia are now targeting broad GPRS rollouts in 2002.

EDGE functionality may be included in GSM/GPRS base stations, and carriers such as Cingular in the United States may use it to boost voice and data capacity on their existing networks; however, new bandwidth has been made available by the FCC in the United States. The costs and timelines for the W-CDMA migration for GSM operators is generally more cautious. In the near to medium term, operators may focus on ensuring the success of their initial GPRS deployments rather than new deployments. The W-CDMA migration may continue to be impacted by increasing capital expenditures (new spectrum and equipment) and some challenges in ensuring high service levels. Network deployment plans in Japan have already been impacted by software and interoperability issues as 3G handsets have been recalled numerous times for software failures.

Voice traffic in wireless networks around the world continues to grow rapidly. It is driven by both an increasing subscriber base and higher usage among existing subscribers. Wireless service providers have traditionally increased capacity by using more spectrum and splitting cell sites by adding additional base stations. Both of these practices, however, have challenges of scale. Spectrum, of course, is finite. Cell splitting, as it adds complexity to network management, also becomes economically impractical at certain levels. Finding new sites for the cell towers is also increasingly difficult, as it can require

Table 2-1

Advantages and disadvantages of 2.5G technology choices

2G Technology	2.5G Upgrade	Description	Advantages	Disadvantages
CMDA-One	CDMA2000 1xRTT	A packet-based technology expected to provide peak data rates up to 307 Kbps (effective rate is expected to be 70 to 144 Kbps) and up to doubling of network voice capacity depending on 1x handset take-up. Capacity improvements are realized as customers transition to new handsets. Uses the same 1.25 MHz channel as CDMA-One.	■ Relatively simple and inexpensive network upgrade, and changes at cell spectrum.	■ Fewer worldwide users and deployment of CDMA networks, limiting regions where users can roam.
			■ Packet-based service enables an always-on connection. Strong effective throughput of around 70 Kbps and up to two times the current voice capacity.	■ Smaller scale may cause handset and network prices to decrease more slowly than for GPRS.
			■ Backward and forward compatible with existing CDMA technology, meaning only customers who want new features must purchase new handsets. Current handsets will continue to function.	■ New handset is needed to make use of data technology advances.

Table 2-1 cont.

Advantages and disadvantages of 2.5G technology choices

2G Technology	2.5G Upgrade	Description	Advantages	Disadvantages
GSM	GPRS	A packet-based technology expected to provide data transfer rates up to 115 Kbps. Effective rates are expected to be around 40 Kbps.	■ Relatively simple network upgrade requiring software upgrade and additional hardware at cell sites. ■ Packet-based service enables an always-on connection and significantly higher data transfer rates than current GSM technology (9.6 Kbps). ■ GSM is the dominant worldwide standard and, accordingly, will drive scale economies in equipment costs.	■ Initial realized data speeds for 2.5G appear lower than CDMA2000 1xRTT. ■ New handset is needed to make use of network advances. Initial handsets have had a variety of performance issues. ■ No interim increase in voice capacity. ■ The next step of W-CDMA requires clean spectrum and two 5 MHz channels.
TDMA	GSM/GPRS	Expected to be followed by most TDMA operators worldwide.	■ Packet-based service enables an always-on connection and significantly higher data transfer rates than current GSM technology (9.6 Kbps).	■ The next step of W-CDMA requires clean spectrum and two 5 MHz channels.

up to 12 months of effort to negotiate zoning regulations. Migrating to next-generation networks should help ease the burden on the network of increased traffic.

Dual and More

When cellular phones were first introduced, all of them were single mode and single band. This offered simplicity to service providers who were focusing most of their efforts on getting cellular services up and running. As more systems got up and running, and as cellular companies started merging and expanding with other companies, multimode and multiband phones became extremely important. U.S. wireless service operators have to supply dual-mode phones that offer CDMA, TDMA, or GSM in addition to 800 MHz analog. Users could use digital features while in their home area and use analog service in areas where digital service was not supported. Overall, this was a viable plan, but it was flawed.

For service operators, operating in analog mode is a waste of resources and a hog of spectrum. Providers can supply several digital calls in the bandwidth of just one analog call. For them, analog calls are very expensive. For customers, analog is not desirable either. Not only is call quality generally worse in analog, but phone battery life is also usually cut by at least half. As a result, the dual-mode/dual-band phone enters the scene. The dual-mode/dual-band phone typically covers 800 MHz analog (AMPS) and a combination of 800 MHz digital and 1,900 MHz digital using either TDMA or CDMA. Some examples of TDMA phones with this technology include the Ericsson KF688, the Mitsubishi T200, and the Motorola ST7797, all of which offer 800 MHz analog, 800 MHz TDMA, and 1,900 MHz TDMA. Recently, many dual-band CDMA phones have also become available, either tri-mode/dual-mode around the world. In the GSM world, the trend is toward using tri-band phones, which cover the European 900 and 1,800 MHz band in addition to the American 1,900 MHz band. Such a phone enables easy roaming between the two countries. Typically, these GAIT phones cover 800 MHz TDMA, 1,900 MHz TDMA, and 1,900 MHz GSM, and models are starting to appear from Nokia,

Ericsson, and Siemens. However, tri-band is certainly not the limit to multiband technology. Recently, quad-band handsets have started to appear, covering the three GSM bands listed previously in addition to the new 850 MHz band, which will be deployed as wireless service operators transition from TDMA to GSM.

WWAN Next-Generation Deployment Update

Next-generation deployments of 2.5G and 3G have been witnessed across the world. The United States uses GPRS and CDMA 1xRTT, Europe deployed GPRS and will soon deploy UMTS, and Asia has released W-CDMA, CDMA 1xRTT, and CDMA 1xEV-DO.

In South Korea, SK Telecom and KT Freetel have already transitioned approximately 40 percent of their subscriber bases to CDMA 1x handsets. Both mobile operators have also experienced an ARPU lift of approximately 10 to 15 percent since launching this service, as subscribers have raised their spending on traditional data services (SMS, ring tones, and screensavers) and have begun paying for games, dating services, and color graphic downloads.

In Japan, NTT DoCoMo's W-CDMA FOMA service has had lackluster success, approaching approximately 112,000 subscribers, due to limited coverage (in Tokyo, Osaka, and Nagoya) and a lack of compelling applications, despite approximately $4 billion of investment to date. KDDI has been more successful; its 'au' CDMA 1xRTT service launched in April 2002 and gained approximately 300,000 subscribers by the end of the first month.

Many mobile operators in Europe launched GPRS service in 2001. These services have not achieved success since the launch because additional desirable content outside of SMS is not provided. Furthermore, no GPRS roaming existed initially, which was a major inhibitor to GPRS adoption.

Verizon and SprintPCS in the United States will have nationwide 1xRTT service by the end of 2002. GPRS services have been launched by Voicestream (now T-Mobile), AT&T Wireless, and Cingular. Figure 2-6 illustrates the technologies that are used around the world.

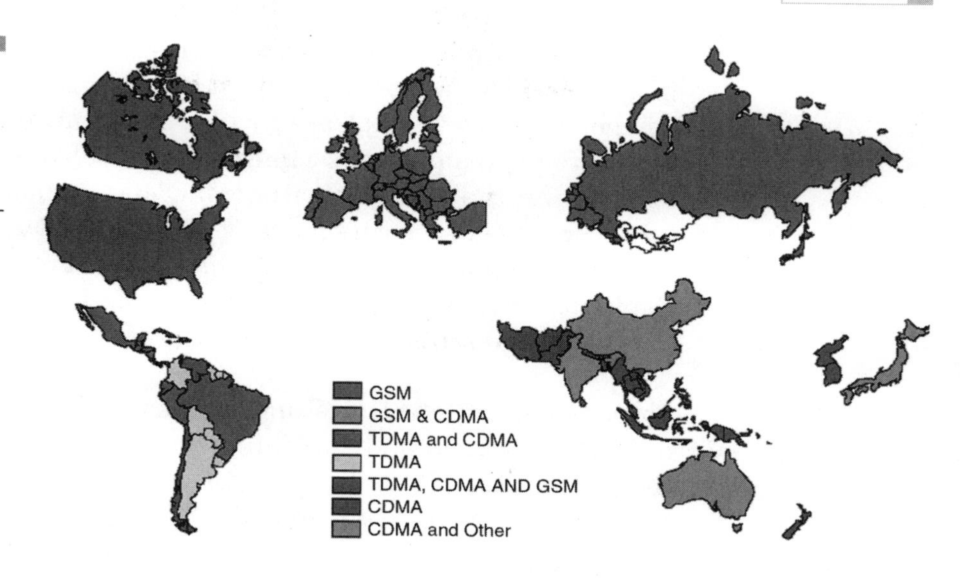

Figure 2-6 Technologies used around the world (Source: Lehman Brothers)

Wireless Data Trends In the past six months, the broad rollout of CDMA 1x in North Korea versus the relatively slow development of W-CDMA in Japan has helped performance and applications on the former to catch up with the rival Japanese technology. So far, the core services remain similar between DoCoMo's FOMA service and the Korean CDMA 1x service—mainly e-mail, songs and/or ring tone downloads, and games. South Korean operators have since enabled camera and motion video functionality over their CDMA 1xRTT networks. Previously, NTT DoCoMo's FOMA seemed more advanced than the Korean services since videoconference and multiaccess functions were available. GPRS networks are much more prevalent throughout Europe and the United States, although throughput speeds are typically half of the next-generation CDMA networks. In-Stat believes that take-up will be driven by SMS/e-mail and character/screen downloads. A small but growing factor will be location-sensitive content such as graphic maps, although such services currently represent a fraction of wireless data revenue in Japan and less than 1 percent of revenue in Korea.

In-Stat also believes that color *liquid crystal display* (LCD) handsets that incorporate digital camera features are compelling devices and should trigger greater data usage as users send and receive images. DoCoMo's survey results of its FOMA service found that 48 percent of users were pleased with its TV/video phone feature. In South Korea, the price of a camera phone could decrease by 30 to 40 percent over the next 12 months from $400 to $600 today, helping to boost the take-up of such services. For traditional GSM markets such as Hong Kong or Europe, photos present an opportunity. However, the limited availability of attractive aesthetics and applications for handsets represents a barrier for adoption.

Costs Although attractive applications are critical for success in mobile data, price and speed remain the key variables for determining adoption. If we compare CDMA 1xRTT and 1xEV-DO with W-CDMA and PDC in Japan and GPRS, we find the following:

- CDMA 1x and W-CDMA transmission speeds are faster and somewhat cheaper from a service price standpoint than GPRS for simple tasks such as SMS, e-mail, and digital photos. Under current plans offered by DoCoMo, SK Telecom, and Hong Kong's CSL, for example, the cost of sending a digital photo amounts to as little as $0.013 on a Japanese W-CDMA program versus $0.046 on a Korean CDMA 1x program and $0.126 on Hong King's GPRS. E-mails are similarly cheaper on W-CDMA and 1xRTT.
- High-volume downloads such as video clips are still expensive ($20 to $100) for all users of W-CDMA, CDMA 1xRTT/1xRTT EV-DO, or GPRS, and take too long for average consumers to tolerate (up to 20 minutes). Until operators further reduce prices, limited demand for such applications will exist. Additionally, it is questionable whether people will be patient enough to wait for these downloads, regardless of the price.
- Color handsets appear to be one of the key success factors for next-generation adoption in Korea and Japan. KT Freetel believes its color LCD users are generating approximately 8 to 10 percent higher ARPU of W36,868 ($29.50) a month versus W33,969 ($27.10) for CDMA 1xRTT black-and-white handset users.

Table 2-2

Next-generation handset costs

	W-CDMA	CDMA 1xRTT	GPRS
Handset prices	\$380–\$510	\$150–\$400	\$150–\$350
Subsidies	\$200–\$300	\$200–\$400	\$50–\$100

Table 2-2 details the next-generation handset price a user can expect to pay today. Note that South Korea has no subsidies.

GPRS remains the cheapest of the three technologies to roll out; however, it offers a relatively limited actual data speed (25 to 45 Kbps) versus the two higher bandwidth services, which typically achieve 50+ Kbps. Given adequate and attractive handsets and applications, GPRS mobile operators may find the service sufficient for the foreseeable future. As GSM and CDMA networks continue to be built out and penetration levels approach 100 percent, these two network technologies will command the majority of worldwide subscribers through 2006.

3G spending on infrastructure will ramp in late 2002, and 3G handsets should ramp some time in mid to late 2003. Success of 2.5G (GPRS) will be critical to 3G. If the data model proves successful and is validated early on, carriers will be more aggressive in their 3G rollouts. Overall, wireless capital expenditures are expected to grow in the low teens. On handsets, the belief is that 2002 could be very difficult due to slowing net addition growth throughout the world and the weak replacement market as subscribers wait for a motivation to upgrade. Given the delays with GPRS, there are not many drivers to spur growth, although these are market characteristics that should be resolving themselves by the end of the year. In Europe, 3G is likely to cost 3.0 to 7.5 billion euros and offer speeds initially in the 64 to 144 Kbps range. As such, high-end multimedia applications will likely disappoint end users, but data rates are more than good enough for the majority of consumer applications, including streaming audio. Actual speed will be partly related to the number of cellular base stations the wireless service operator installs as well as the number of people accessing the network at a given point in time. It is estimated that an upgrade to 2.5G costs less than $10 per subscriber compared to 16 times that amount for a new infrastructure deployment. Given this difference in cost, Bear Stearns has done a rough estimate of the ROI on both 2.5G and 3G builds, which is provided in Table 2-3.

Using assumptions about the life of the technology, the cost per geography deployed, the number of networks to be built, the number

Table 2-3

GPRS versus 3G costs

	GPRS	3G
Rollouts	2001	2003
Active period	2002–2003	2004–2010
Cost per geography ($ in millions)	300	7,000
Number of upgrades per geography	5	5
Number of geographies[1]	6	6
Total network cost ($ in millions)[2]	9,000	210,000
Population of western Europe (millions)	240	250
Wireless subscription penetration rate	70%	85%
Number of wireless subscribers (millions)	168	213
Average take rate	20%	50%
Average number of data subs	34	106
ARPU increase per month	15	25
Life of network (months)	24	84
Gross revenues	12,096	223,125
Internal rate of return (IRR)	34.7%	1.7%

[1]Iberia, Benelux, France, Italy, Germany, and the United Kingdom

[2]Spectrum license is a sunk cost.

Source: Bear Stearns

of subscribers to adopt the services over their life, and the ARPU increase from those services, Bear Stearns estimated that GPRS has an extremely favorable projected IRR (34.7 percent) and the 3G's IRR is a very challenging 1.7 percent. Although changes in assumptions could alter these calculations, they are likely to move in step for each technology. Given these results and Bear Stern's projections, 3G technology in its first iteration is unlikely to offer more compelling end-user capabilities than 2.5G. In addition, as one carrier in a given market builds and deploys a 3G network, it will force other carriers to build them to keep from being at a competitive disadvantage, even if those build-outs do not appear to be economical when taken as a whole.

3G is currently a case where economic realities conflict with what can be accomplished from a technological perspective. 2.5G networks provide 50 to 80 Kbps and cost 20 percent of a 3G network build. Functionality will encompass everything except for high-quality streaming, an incremental capability that does not make economic sense to deploy for the benefit derived thereof. One way to bring down those price points would be through consolidation so that a fewer number of 3G networks would be built in a given region.

3G Network Sharing The cost of 3G licenses has put many mobile operators in debt. Regulators are searching for alternatives to assist with 3G deployments to improve these companies' financial well being. One solution would require the operators to share networks (multivendor base stations or other kinds of equipment), which should not represent a major technical hurdle for vendors or require extensive development. For example, Nokia has developed a multioperator W-CDMA base station, which enables up to four operators to share the same base station and operate their frequencies independently. Network sharing could also alleviate the debt burden for some frail customers and may also provide some relief for vendor financing commitments.

The German regulator (RegTP) recently softened the rules regarding 3G network sharing. Operators can now share base stations and BSCs in addition to sites, masts, cabling, and combiners. The purpose is to enable operators to offer dedicated services over their own licensed frequencies while alleviating their debt burden.

The trend toward network sharing may not necessarily lead to significantly reduced opportunities for vendors for the following reasons:

- Network sharing appears to be limited to a very small number of countries at this stage (Germany and the United Kingdom).
- Shared base station cities must be developed. Urban implementation is more difficult as traffic will be much denser and QoS would need to be monitored carefully.

The key objective for operators building a 3G network is most likely to ensure the highest QoS. Thus, network sharing should be regarded as a temporary cost-saving measure of deployment. As an approximate estimate, network sharing could help operators realize savings of approximately 15 to 30 percent of their capital expenditure budget.

Due to all the issues regarding 3G deployments, it is important for operators to investigate all the technologies that are available to them in order to create a smart economic decision.

Bluetooth

Bluetooth is a specification for short-range wireless communications that operates in the 2.4 GHz *Industrial, Scientific, and Medical* (ISM) band. Bluetooth supports distances of 10 meters, which can be extended to 100 meters by using an amplifier. The standard supports both voice and data communications—data can only be supported at speeds up to 721 Kbps. Bluetooth aims to eliminate the need for wires or cables and create a wireless *personal area network* (PAN). A WPAN consists of the connections between the devices surrounding an individual such as *personal digital assistants* (PDAs), mobile phones, and printers. Bluetooth applications include wireless headsets for mobile phones and the ability to access a printer from a PDA. They also offer the possibility of building airport networks that provide fast check-in via Bluetooth-enabled devices within range of the ticket counter. In addition, Bluetooth could enable vending-machine

Figure 2-7 Bluetooth usage diagram (Source: Bluetooth Association)

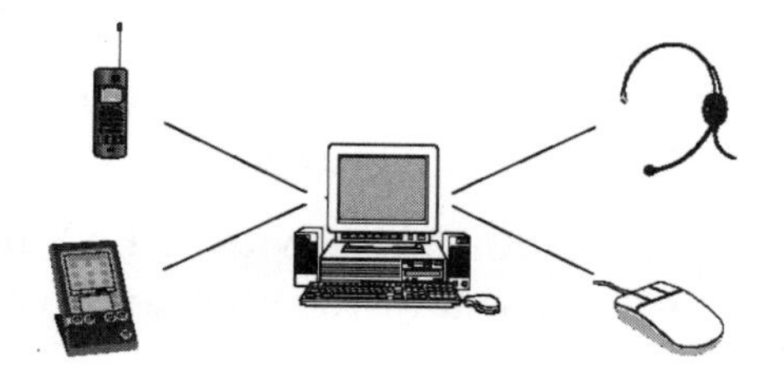

purchases using mobile phones. Going forward, Bluetooth may also provide corporate connectivity to corporate intranets and the Internet for mobile devices from access points strategically placed throughout a company. Figure 2-7 illustrates how Bluetooth can function as a WPAN.

Bluetooth has the following benefits:

- Bluetooth uses *frequency-hopping spread spectrum* (FHSS) and has a hop rate of 1,600 *hops per second* (hps). Should Bluetooth experience interference on one frequency, the problem will be cleared after a hop to the next one.
- Bluetooth supports both voice and data. Voice is supported using a 64 Kbps data rate. Pure data is supported in two ways: asymmetrically and symmetrically. Asymmetric data runs at 721 Kbps in one direction and 57.6 Kbps in the return direction, whereas symmetric data runs at 432.6 Kbps in both directions.
- Up to eight devices can operate in one piconet. (The total throughput is shared among users.)
- Bluetooth has built-in security through authentication and encryption in the baseband protocol.
- The technology supports both isochronous (simultaneous) and asynchronous services.

Bluetooth Use in Cellular Handsets

Bluetooth was conceived as a means of connecting many different pieces of equipment for many different purposes. As a result, many software stacks have been developed. Within handsets, Bluetooth's uses are more limited, at least initially. These uses include the following:

- Hands-free voice communication between a handset and a hands-free device such as a headset or automobile speakerphone.
- Data communications between a handset and a desktop computer or laptop to synchronize a handset's calendar or phonebook with that of the computer's.
- Data communications between a handset and a laptop, giving the laptop access to data services provided by the handset.
- Communications between a handset or computer and printer, enabling the handset user to print out calendars and phonebooks from the handset.
- Short-range communications from an access point to a handset, enabling a user to check flight reservations, get customer service, and make purchases from merchants offering this service.

Bluetooth Deployment Update

Mobile operators have not focused on public Bluetooth a great deal up to now. Remember that the primary purpose of Bluetooth is to replace a cord, not to provide high-speed network access. However, one public Bluetooth example does exist today. Red-M and British Telecom are working together to support the British Rail. The rail service is planning to roll out a Bluetooth-based wireless access network for British Rail passengers. From a terminal or train on a main route, a range of services such as Internet content, e-mail access, travel information, and e-ticketing will be offered.

WLANs

WLAN technologies are designed to transmit and receive data over the air, minimizing the need for wired communications. WLANs may either act as a substitute for a wired LAN or as an extension of an existing LAN. WLANs fill many of the same purposes and provide many of the same benefits of wired LANs. WLANs have the added benefits of reduced costs related to the elimination of unnecessary cables, roaming, ease of installation, and extended range. The WLAN sector encompasses several different standards including 802.11b, 802.11a, *Home Radio Frequency* (HomeRF), and *high-performance radio local area network* 2 (HiperLAN2).

WLAN Bands and Technologies

WLANs may be based on several types of technology and work over various frequency bands. First, 900 MHz systems became available, but these products never achieved very high data rates and most that are still in existence are legacy products. Next, the 2.4 GHz products became available. These offered faster data rates in what was, at the time, a relatively unused band. Today, the vast majority of WLAN products use the 2.4 GHz band. Although the 5 GHz 802.11a standard was approved at the same time as the 2.4 GHz 802.11b standard, it has taken much longer for the higher frequency (5 GHz) products to appear.

The 2.4 GHz Band The 2.4 GHz band occupies 2.400 to 2.4805 GHz in the United States and Canada and provides a bandwidth of 80.5 MHz. When WLAN products first appeared in this band, standards for transmission protocols had not yet been adopted and WLAN *network interface cards* (NICs) from one manufacturer would not work with WLAN NICs from another manufacturer. Because of this, early WLAN growth was slow and early adopters became disillusioned with the technology.

Seeing this as a problem, the *Institute of Electrical and Electronics Engineers* (IEEE) created a WLAN standard so that WLAN products from different manufacturers could interoperate. After much work and debate, the IEEE ratified the 802.11 standard in 1997, which specified a single *Media Access Control* (MAC) sublayer and three *physical* (PHY) layer specifications. For the first time, WLAN products from one manufacturer would be able to interoperate with those from another (assuming that both used the same PHY layer).

Although this was an enormous step in the right direction, the approval of three PHY layers was cause for concern. Many manufacturers chose FHSS as their PHY layer, whereas others, such as Lucent (now Agere and recently acquired by Proxim), chose *direct sequence spread spectrum* (DSSS). (802.11 also specifies *infrared* [IR] as a PHY layer, but very few manufacturers build products to this specification because of the line-of-sight limitations.) Because of this, interoperability problems persisted.

Not wanting to repeat past mistakes, in September 1999, the IEEE finalized 802.11b, the high-rate extension of the 802.11 standard. 802.11b was the standard that the market was waiting for, and sales of WLAN products took off. By the end of 2000, only a little more than a year after 802.11b was ratified, more than 30 companies were selling 802.11b products—for example, Cisco, Nokia, and Proxim. Although 802.11b has become the most widely embraced standard in WLAN history, the IEEE also realizes that progress must continue, and Task Group G was formed to create a higher rate successor to 802.11b.

The 5 GHz Band The initial 5 GHz band in the United States, which was authorized by the FCC in 1985, occupies the band from 5.725 to 5.825 GHz and provides 100 MHz of bandwidth. In 1997, the FCC authorized an additional 200 MHz of spectrum by adding the 5.15 to 5.35 GHz band.

Generally, 200 mW in the 5 GHz band is considered to give an equivalent range to 100 mW in the 2.4 GHz band. This means that radios operating in the U.S. 5.15 to 5.35 GHz band will have somewhat less range than a typical 802.11b radio, whereas a radio operating at the 200 mW limit in the 5.25 to 5.35 GHz band will have a range that is approximately the same as a 2.4 GHz DSSS radio. The 1 W, 5.725 to 5.878 GHz band will typically be used for high-power point-to-point bridges.

Multiple Standards Could Cause Interference

A common concern for all unlicensed bands is interference between the devices using the spectrum. This is a serious concern among unlicensed bands, such as 49, 330, 900, and 2.4 GHz. The 2.4 and 5 GHz bands are subject to overcrowding and interference. As more standards and wireless systems are introduced to both ISM bands, the industry is concerned the performance might be greatly affected. Proximity is another major factor to take into consideration in terms of interference. Interference between systems only occurs when the systems are extremely close to each other.

Certainly, 802.11x is the technology in the driver's seat, with the current WiFi (802.11b) products bringing speeds that are enough to catch people's attention in the home. A variety of wireless technologies are listed and explained in the following sections. Many pertain to the family of 802.11x technologies. 802.11x technologies, including 802.11b, 802.11g, and 802.11a, hold great promise for true high-speed (+10 Mbps) networking in the home. However, other wireless technologies that are compelling for the home are available, such as *Ultra Wideband* (UWB) and Wireless 1394. Ad-hoc peer-to-peer mesh technology is quite compelling for wide area and home networking, although its esoteric nature makes the technology's message difficult to evangelize.

IEEE 802.11

IEEE 802.xx is a set of specifications for LANs from the IEEE. Most wired networks conform to 802.3, the specification for Ethernet networks based on *carrier sense multiple access with collision detection* (CSMA/CD), or 802.5, the specification for token ring networks. 802.11 defines the standard for WLANs encompassing three incompatible (noninteroperable) technologies: FHSS, DSSS, and IR. This standard, which was ratified in 1997, set guidelines for interoperability for products in each specific technology category. These standards apply to products with speeds up to 2 Mbps.

802.11x is commonly used as a generic term for the family of higher-speed versions of the original IEEE 802.11, including 802.11b, 802.11a, and 802.11g, which all share the same MAC.

802.11b In June 1997, the IEEE finalized the initial standard for WLANs—IEEE 802.11. This standard specifies a 2.4 GHz operating frequency with data rates of 1 and 2 Mbps using either DSSS or FHSS. The IEEE 802.11 Working Group has since published two supplements to this 802.11 standard: 802.11b (direct sequence in the 2.4 GHz band) and 802.11a (*orthogonal frequency division multiplexing* [OFDM] in the 5 GHz band).

IEEE 802.11b was ratified in September 1999 and is getting much attention. This standard essentially legitimizes Ethernet-like speeds for WLANs. The IEEE 802.11b high-rate standard defines interoperability standards for products employing DSSS technology, operating at speeds up to 11 Mbps. Supporters of this standard are largely the members of the *Wireless Ethernet Compatibility Alliance* (WECA). This organization is made up of members from the networking equipment, wireless, consumer electronics, and silicon component industries. IEEE 802.11b high-rate products started shipping in late 1999 (Aironet released its 11 Mbps products in late 1998 before the standard was actually ratified).

The IEEE 802.11 Working Group currently has several task groups working on MAC enhancements: Task Group H (interaccess point protocol), Task Group I (security), Task Group E (QoS), and Task Group G (+20 Mbps speeds in the 2.4 GHz band). Additionally, Task Group H is working on adding *Dynamic Frequency Selection* (DFS) and *Transmit Power Control* (TPC) features to 802.11a in order for the technology to gain regulatory approval in Europe.

802.11g IEEE 802.11 Task Group G was charged with coming up with a draft for IEEE 802.11g, which is basically the higher-speed extension to IEEE 802.11b. General requirements for any proposal for 802.11g included the following:

- The proposal must be an extension of IEEE 802.11b.
- The proposal must specify a PHY layer that implements all mandatory portions of the IEEE 802.11b PHY standard.
- It must comply with IEEE patent policy (the company formulating the proposal must be willing to provide licenses at minimal cost).
- The proposal must be backwards compatible with IEEE 802.11b.

As of March 2001, two proposals remained in the mix: Texas Instruments' *packet binary convolutional coding* (PBCC) and Intersil's OFDM. A heated political, as well as technological, debate ensued over the two. In May 2001, one proposal remained—Intersil's OFDM. Intersil's OFDM did not receive the necessary 75 percent acceptance rate at the July 2001 meeting, and the September 2001 meeting was cancelled due to the September 11th terrorist attacks. At the July meeting, Texas Instruments representatives submitted a compromise proposal calling for a requirement of both OFDM/*complementary code keying* (CCK) and PBCC transmitters in a last ditch effort to get PBCC into 802.11g. At the November meeting, a compromise was made to make Intersil's OFDM the basis of 802.11g and make Texas Instruments' PBCC optional. 802.11g is a promising technology for the following reasons:

- 802.11g products will provide backwards compatibility with 802.11b products, significantly decreasing the obsolescence paranoia among the large and growing installed base of 802.11b end users.
- 802.11g, operating in the 2.4 GHz band, is legal under the regulatory laws of ETSI, while 5 GHz 802.11a technology is currently illegal in Europe.
- The use of OFDM as the core PHY technology within both 802.11g and 802.11a will make the development of combo 2.4 and 5 GHz solutions less complicated, cheaper, and presumably more efficient than the originally planned dual 802.11b/802.11a solutions. 802.11b uses DSSS technology in the PHY layer, whereas 802.11a uses OFDM.

Although 802.11g has strong support from Intersil, which provides approximately 65 percent of the WLAN integrated circuit chipset shipments in the 802.11b market, the success of 802.11g will depend on an embrace of the technology by the likes of Cisco, 3Com, Agere, and/or other trusted vendors in the networking space.

Penetration of 802.11g will ultimately be driven by the overall performance reports on the initial 5 GHz products released in 2002.

802.11e Task Group E of the IEEE 802.11 Working Group now focuses on QoS enhancements to IEEE 802.11. Task Group E used to encompass both QoS and security, but, with the scope of the issues and the different factions involved, the decision was made to spin off a separate task group for security: Task Group I. Sharewave is very active in this task group, as its Whitecap protocol has been designed to serve as a proprietary QoS enhancement to 11 Mbps products. Netgear and Panasonic, for example, sell 802.11b-like products in retail with Sharewave's Whitecap protocol in an effort to boost QoS for the home. Task Group E is the primary group within the IEEE 802.11 Working Group and coordinates with the Wireless 1394 Task Group.

802.11i Task Group I is devoted to improving the much criticized security standard of 802.11 products. This is a high-profile, high-pressure task group. End users, industry pundits, and so on are all waiting for the new standard that this group will be draft. The goal was to have a standard ratified by the first half of 2002. In 2000, Task Group E was formed to address both security and QoS. In 2001, two different task groups were formed from the original Task Group E: Task Group G to focus on QoS and Task Group I to focus on security. WLAN proponents hope to have this critical security standard ratified by 2002 in order for 802.11i-compliant end products to enter the market by 2003.

Some of the proposals being examined by Task Group I include authentication mechanisms such as *Remote Access Dial-In User Service* (RADIUS) and Kerberos; IEEE 802.1x, a draft standard for port-based network access control that would provide for authenticated network access to 802.11 wireless networks; and *Advanced Encryption Standard* (AES), a highly touted encryption standard considered by many to be the predecessor to *Data Encryption Standard* [DES] and *Triple DES* (3DES).

802.11f Task Group F of the 802.11 Working Group has the inter-access point protocol standard as its task. The main goal of this standard is to provide an increasing level of vendor access point interoperability. In many cases, access points from multiple vendors must be stripped down to their bare minimum features in order to interoperate in a network.

802.11h Task Group H's goal is to add features to the current 802.11 5 GHz standard so regulatory bodies outside of the United States will allow 5 GHz 802.11 products. Currently, under ETSI's rules and regulations, IEEE 802.11a products are illegal in Europe due to spectrum allocation issues. The purpose of Task Group H is to provide the IEEE 802.11 MAC and PHY layers network management and control extensions for spectrum and transmit power management in 5 GHz license-exempt band. It also plans to provide improvements in channel energy measurement and reporting, channel coverage in many regulatory domains, and DFS and TPC mechanisms.

802.11a IEEE 802.11a was ratified in September 1999 in conjunction with the ratification of 802.11b. This standard employs an OFDM PHY layer for data transmission. Products will operate in the 5 GHz band with speeds up to 54 Mbps. The IEEE 802.11a standard specifies an OFDM PHY layer that splits an information signal across 52 separate subcarriers to transmit data at a rate of 6, 9, 12, 18, 24, 36, 48, or 54 Mbps. The 6, 12, and 24 Mbps data rates are mandatory for all products. Four of the subcarriers are pilot subcarriers that the system uses as a reference to disregard frequency or phase shifts of the signal during transmission.

Wideband OFDM (W-OFDM) is a variation of OFDM and is the basis of the IEEE 802.11a standard. W-OFDM differs from OFDM in that the spacing between carriers in W-OFDM is large enough that any frequency errors between the transmitter and receiver are only a small fraction of the spacing; therefore, they have a negligible effect on system performance. W-OFDM is very different from DSSS or FHSS. Both of these types of spread spectrum broadcast over a wide range of frequencies while attempting to overcome noise and multipath errors.

WWAN, WLAN, and Bluetooth Technology Comparison

Now that the three different technologies have been discussed, Tables 2-4 and 2-5 compare the key characteristics head to head.

Network Performance Compared

Table 2-6 is Morgan Stanley's comparative data throughput analysis of the preceding network types.

Now that different technologies have been discussed, it is important to understand what level of performance each can produce. Although 2.5G and 3G network technologies offer much greater data throughput speeds, the geographic coverage (most likely found in urban areas) of these networks compared to the 1G and 2G will be substantially less, primarily because of the associated costs on implementing each base station.

Lack of Handset Penetration

Currently, the average user replaces his or her handset between 12 months and 2 years. However, the network operators seeking to grow their subscriber base at almost any cost artificially generated this churn. This has led to the widespread subsidy of handsets to users and made it more beneficial to swap networks and gain a new handset than to remain with the current network. Recently, following the stock market decline and the widely publicized debt overhead of the wireless service operators, these subsidies are now being reduced and operators are seeking to reduce churn and move their customer base away from a prepay mentality and back to a contract or postpaid base. With this in mind, handset churn is likely to decrease since the cost of handset will generally increase, creating a potential barrier to the adoption of new handset technology.

Table 2-4

WLAN and Bluetooth network characteristics

WLAN Technologies	Technology Detail	Range	Data Throughput (Forward Link)	Availability
IEEE 802.11	Sanctioned by the IEEE to run on a single MAC for three physical layers: FHSS, DSSS, and IR. Includes inherent data encryption.	Range is approximately 200 to 300 feet.	Standards for 2.4 GHz radios running at speeds up to 2 Mbps.	Standard was ratified in 1997. These products are continually losing traction against 802.11b products.
IEEE 802.11b	DSSS technology. MAC enhancements in standard are being worked on, including security and QoS features.	Range is approximately 200 to 300 feet.	2.4 GHz supports speeds up to 11 Mbps.	Standard was ratified in September 1999. Volume shipments began in late 1999.
IEEE 802.11g	Intersil's OFDM mandatory, whereas Texas Instruments' PBCC is optional. The final standard is not expected until late 2002 or early 2003.	Range is expected to be between that of 802.11b and 802.11a—approximately 150 to 250 feet. Since it is still operating in the 2.4 GHz band as is 802.11b, range doesn't deteriorate as fast as if it moved to 5 GHz.	2.4 GHz supports speeds up to 54 Mbps.	Chipsets from Intersil was available in 2002. Texas Instruments was expected to release a chipset incorporating CCK/OFDM with PBCC in 2002. Both companies expected to have chipsets sampling before the final 802.11g standard is ratified.
IEEE 802.11a	Employs OFDM technology. Single-carrier frequency selection. Connectionless. Differs from HiperLAN2 only in the MAC layer.	Range is expected to be less than half that of 2.4 GHz products—approximately 50 to 100 feet.	5 GHz supports speeds up to 54 Mbps.	Hardware shipments began in very small volumes in the fourth quarter of 2001 and have continued to ramp up throughout 2002.

(continued)

Table 2-4 cont.

WLAN and Bluetooth network characteristics

WLAN Technologies	Technology Detail	Range	Data Throughput (Forward Link)	Availability
HiperLAN2	OFDM technology. Single carrier with DFS. Connection oriented.	Range is expected to be less than half that of 2.4 GHz products—approximately 50 to 100 feet.	5 GHz supports speeds up to 54 Mbps.	Standard is rapidly losing its clout against WiFi5 (IEEE 802.11a). WiFi5 products are expected to be generally accepted by Europe with the addition of TPC and DFS.
HomeRF	TDMA for isochronous (for example, voice); CSMA/CA for high-speed data. Frequency-hopping technology with voice support based on the DECT standard.	Range is approximately 200 to 300 feet.	1 and 2 Mbps modes.	Products were shipped in 2000.
Wideband frequency hopping (WBFH) (HomeRF 2.0)	Frequency-hopping technology with voice support based on the DECT standard. Will be backwards compatible with current HomeRF products.	Range is approximately 200 to 300 feet.	2.4 GHz supports speeds up to 10 Mbps.	WBFH was approved by FCC in the summer of 2000. Proxim released products in 2001.
Bluetooth	Uses a quick, frequency-hopping (1,600 hps), packet-switched protocol. 2.4 GHz; 1 mW Tx; low power consumption and small form factors are possible. Viewed as a PAN technology.	Range is approximately 10 meters. Generally viewed as a cable replacement technology in comparison to 802.11b and 802.11a.	721 Kbps.	Bluetooth-enabled end products began to ship in 2001.

Source: In-Stat/MDR, May 2002

Table 2-5

WWAN network characteristics

WLAN Technologies	Technology Detail	Range	Data Throughput (Forward Link)	Availability
CDMA2000[3]	This is a superset of advanced CDMA technologies, but not one technology in and of itself.	≈14 miles	Rates up to 14.4 Kbps, average 9 to 14 Kbps.	Deployed in the United States. Approximately 3.5 million data users.
CDMA2000 1xRTT[3]	A higher-capacity, higher-efficiency modulation that is backwards compatible with existing IS-95 equipment.	≈14 miles	Rates up to 144 Kbps are possible, but the reality will be 50 to 120 Kbps in most cases. Phase 2 will support data rates up to 307 Kbps.	In excess of 8 million subscribers currently in South Korea. Is currently being deployed in the United States.
CDMA2000 1x+ A22EV-DO[3]	A DO modulation employing core CDMA technology.	≈14 miles	Data rates of up to 2.4 Mbps are possible. Typical data rates are 0.8 to 1.4 Mbps	Available in South Korea currently. It's in trials in United States. May be commercially available in the United States in early 2003.
CDMA2000 1xEV-DV[3]	A DV modulation also based on core CDMA technology.	≈14 miles	All the details have yet to be worked out, but it's expected that EV-DV data rates will be similar to EV-DO data rates. Greater than of 2 Mbps is possible.	2005 to 2006 range.
W-CDMA	A CDMA-type technology employing a (5 MHz) carrier.	≈14 miles	Data rates up to 384 Kbps are possible, but higher speeds will eventually be possible. Average speeds are 80 to 90 Kbps.	NTT DoCoMo in Japan started commercial service in October 2001.

(continued)

Table 2-5 cont.

WWAN network characteristics

WLAN Technologies	Technology Detail	Range	Data Throughput (Forward Link)	Availability
GPRS	An evolutionary step from GSM that enables multiple time slices to be chained together, providing higher data speeds.	≈14 miles	There are several classes, or services, available; these vary depending on resource allocation. Typically, data rates are in the 30 to 45 Kbps range in actual practice.	It is available now in Europe and in the United States.
EDGE	An evolutionary step of GPRS that provides for higher data rates through a more advanced modulation scheme.	≈14 miles	Data rates should be 2 to 3 times faster than GPRS, but like GPRS, different classes of service will be offered. Theoretical data speeds can be up to 384 Kbps.	Cingular and AT&T hope to offer service in the United States in early 2003. At this point, Europe is still an unknown.

[3]There is really not a technology that is just CDMA2000. Rather, CDMA2000 encompasses a series of technological advances including 1x+A14RTT, EV-DO, and EV-DV.

Source: In-Stat/MDR, June 2002

Table 2-6

Network technology performance compared

		MP3 File	Excel File	Digital Photo	E-mail	Web Page
	File Size (Kb)	**4,000**	**2,000**	**100**	**25**	**75**
	Number Downloaded	**8**	**1**	**12**	**12**	**1**
	Data Dump (Kb)	**32,000**	**2,000**	**1,200**	**300**	**75**
Kbps	**Description**	**Downloaded Time in Minutes (or Seconds)**				
9.6	1G circuit switched	444 (26,667)	28 (1.667)	17 (1,000)	4.2 (250.0)	1.0 (62.5)
14.4	1G circuit switched	296 (17,778)	19 (1,111)	11/666.7	2.8 (166.7)	0.7 (41.7)
19.2	2G packet data	222 (13,333)	14 (833.3)	8.3 (500.0)	2.1 (125.0)	0.5 (31.3)
56.0	2.5G packet data (GPRS)	76 (4,571)	4.8 (285.7)	2.9 (171.4)	0.7 (42.9)	0.2 (10.7)
144.0	2.5G packet data (CDMA 1x)	30 (1,778)	1.9 (111.1)	1.1 (66.7)	0.3 (16.7)	0.1 (4.2)
384.0	3G packet data—mobile	11 (666.7)	0.7 (41.7)	0.4 (25.0)	0.1 (6.3)	0.0 (1.6)
1,000.0	Bluetooth	4.3 (256.0)	0.3 (16.0)	0.2 (9.6)	0.0 (2.4)	0.0 (0.6)
2,000.0	3G packet data—stationary	2.1 (128.0)	0.1 (8.0)	0.1 (4.8)	0.0 (1.2)	0.0 (0.3)
11,000.0	WLAN	0.4 (23.3)	0.0 (1.5)	0.0 (0.9)	0.0 (0.2)	0.0 (0.1)

Source: Morgan Stanley

Handset Shipment Forecasts

Although we remain positive on the potential of MMS to become a successful evolution of SMS, the timing of consumers' take-up of the service will be key. Until MMS becomes a widely used service, purchases of *Multimedia Message Centers* (MMSCs) will be limited and predominantly small, low-capacity systems.

Two factors that will play a big role in the timing issue will be the rollout of 2.5G and 3G networks and, even more importantly, the availability and penetration of handsets with MMS functionality.

Figure 2-8 shows the global handset shipment forecasts breakout between new and replacement from Merrill Lynch.

However, the concern surrounding the handset implications for EMS and MMS lies in the percentage and the timing of the rollout of handsets that support EMS and MMS functionality within the overall shipment forecasts.

Table 2-7 presents out the estimates for the percentage of the total shipment forecasts that relate first to 2.5G- and 3G-compatible handsets and then to EMS and MMS handsets globally. The total of EMS-capable handsets include those that work on proprietary enhanced messaging solutions such as Nokia.

Of the global forecasts, Merrill estimates that approximately 70 percent relate to Asia (including Japan) and Europe. Despite a

Figure 2-8
Global handset shipment forecasts (Source: Merrill Lynch)

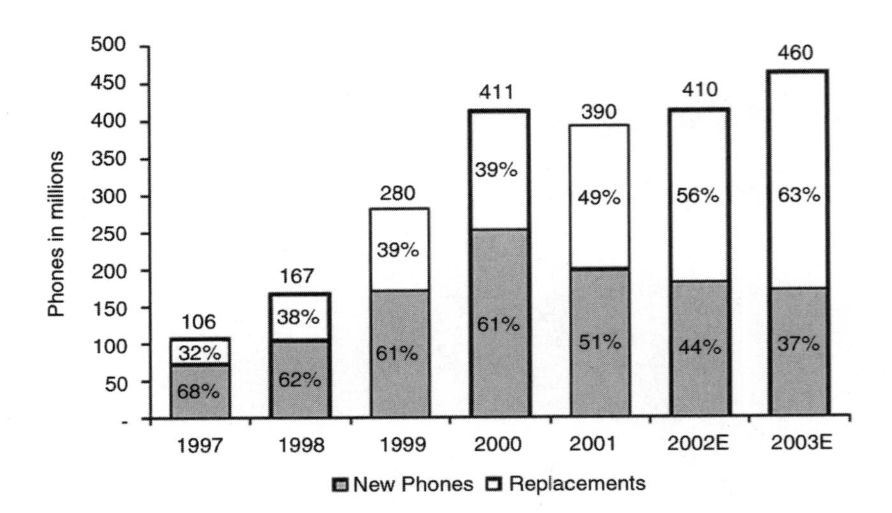

Table 2-7

Percentage of EMS and MMS capable handsets shipped

In Millions	1999	2000	2001	2002	2003
2.5G and 3G phones	NA	NA	15	74	206
% with EMS functionality	NA	NA	30%	45%	35%
Phones with EMS functionality	NA	NA	5	33	72
% with MMS functionality	NA	NA	5%	30%	50%
Phones with MMS functionality	NA	NA	1	22	103
% of total shipments of phones with EMS functionality	0%	0%	1%	8%	16%
% of total shipments of phones with MMS functionality	0%	0%	0%	5%	22%

Source: Merrill Lynch

pickup in the volumes shipped in 2002, the analyst firm believes real volumes will be seen in 2003.

In order for the enhanced P2P messaging market to take off, a sufficient population of users is required. It is estimated by In-Stat and Morgan Stanley that penetration of capable devices needs to achieve approximately 25 percent before significant usage can be achieved. Hence, the timing of the rollout of handsets should witness MMS volumes from 2003 onwards and significant usage one to two years later.

The cost of handsets will also be a key issue. In recent presentations, Vodafone highlighted that 2G terminals now cost 50 to 100 Euros, GPRS terminals cost 80 to 250 Euros, and 3G terminals cost 300 to 600 Euros. It is likely to take the normal evolution time of two to three years for enhanced handset prices to come into line with current handset prices.

It should be noted that the MMSC can also play an important role in providing access to other platforms such as *Unified Messaging* (UM), e-mail, and the Internet. Therefore, once compliant handsets are available and these services are up and running for the mobile environment, it can be argued that users will be able to take advantage of MMS immediately even if no significant population of users exists. However, it is unlikely that such services will be available immediately (particularly without startup problems) and also unlikely that they will be at a cost that is attractive to many users in the early stages. In addition, the availability and the quality of such services are likely to be impacted by the commercial availability of high-speed networks. The ongoing delays and pushback in the roll-out of 3G networks will, therefore, be an issue.

3G Driving Applications

In general, 3G applications refer to those that will be functional only at broadband access speeds or higher. Therefore, this refers almost exclusively to streaming audio and video. Our belief is that the evolution of mobile computing will see devices moving in and out of different networks, whether they are WPANs, WLANs, or WWANs. In addition, looking at WPAN and WLAN technologies, whether they are Bluetooth or 802.11b, little evidence exists to refute the notion that they are both ahead of 3G in terms of delivering cost-effective high-speed connectivity to devices. Combine this with ever decreasing prices for storage and the vision for the evolution of multimedia applications on handheld devices is one of devices moving in and out of fast access areas where desired content is stored on a device, which can then be played on the user's timetable. Although this does not completely eliminate the transmission of multimedia material from WWAN networks, it will limit its role to that of filling in the gaps rather than being the primary transport mechanism for content. Not only does this vision bring about cheaper deployment of high-speed access, but it also reduces the problem of much of the spectrum shortage that wireless service operators confront today, particularly in the United States and some Asian/Pacific countries.

After investigating the WLAN technologies such as 802.11a, which will bump access speeds to 54 Mbps, it will be difficult to believe that WWANs will ever be able to bridge the price/performance gap of WLANs. The value in WWANs is not speed, but coverage and mobility. Users do not pay for great service; they pay for mobility.

This is not to say that it is not possible that any broadband applications will be delivered over WWANs; it only states that they will have to be more value added from the consumer's perspective. For example, applications such as interactive gaming and watching live sporting events are expected to be features delivered over WWANs. However, these are applications that consumers have shown a willingness to pay for and, as such, can justify the greater expense associated with delivering them over WWANs. The point here is simply that WWANs will be the predominant mechanism for the delivery of all multimedia content to handheld devices. But WLANs and WPANs will come into play.

Compression technology is at the heart of enabling multimedia applications on less bandwidth. The transfer format for *Moving Pictures Experts Group 4* (MPEG-4) digital video has been agreed upon by all three main technical bodies: the *Internet Engineering Task Force* (IETF), the ITU, and the *Third-Generation Partnership Project* (3GPP). The standardization is important if video-based calls are to be made across telephones, mobile phones, and PCs that traverse different networks. The standard is based on the *User Datagram Protocol* (UDP) rather than the *Transmission Control Protocol/IP* (TCP/IP) as it is less data intensive. The drawback is that there is no mechanism for recovering lost packets. Companies such as PacketVideo and Solid Streaming utilize MPEG-4 at the core of their video-enablement technologies and then modify delivery technology to further deal with bandwidth and device constraints.

As compression technologies continue to enable the delivery of multimedia, applications will become a more prolific part of a wireless service provider's repertoire of services. Figure 2-9 illustrates Bytemobile's compression result on a GPRS network.

An example of a 3G application has already been offered by Freeserve in the United Kingdom. For about $1.60 per day, which a user pays through an SMS message, a fan gains access to two six-minute clips of Lennox Lewis' preparations for his fight with Hasim "The Rock" Rahman. This provides fans with access that was previously unattainable, including prefight preparations and seeing who visits the boxer before the bout. Although this behind-the-scene footage today is delivered via a PC and paid for via SMS and Freeserve's relationship with Vodafone, it demonstrates three key features on the 3G future of multimedia content.

Figure 2-9 Compression effects (Source: Bytemobile)

2X to 7X speed-up

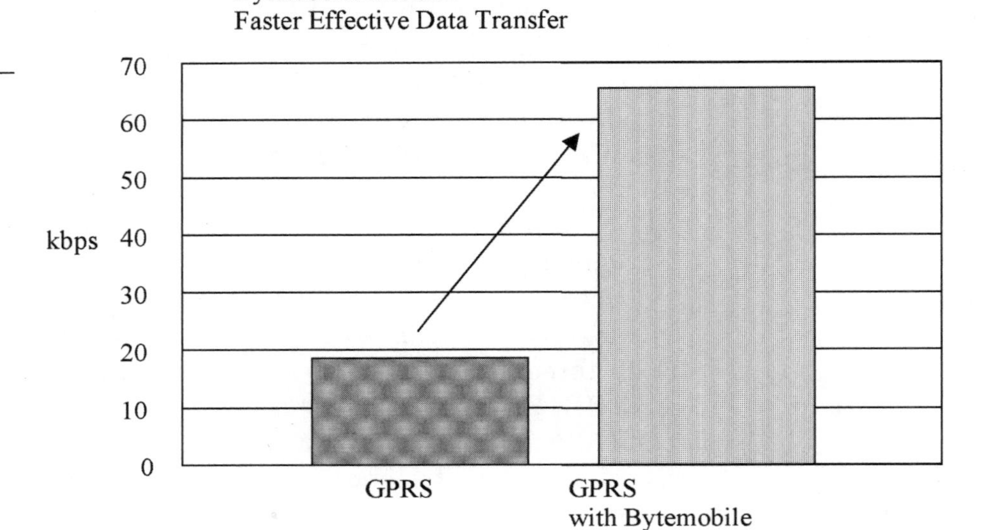

6X to 10X reduction in file size, allowing retail price reduction

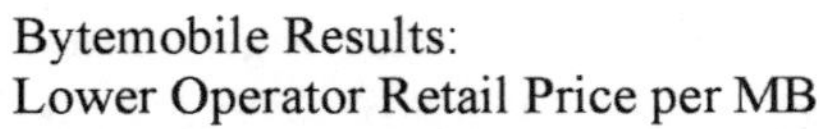

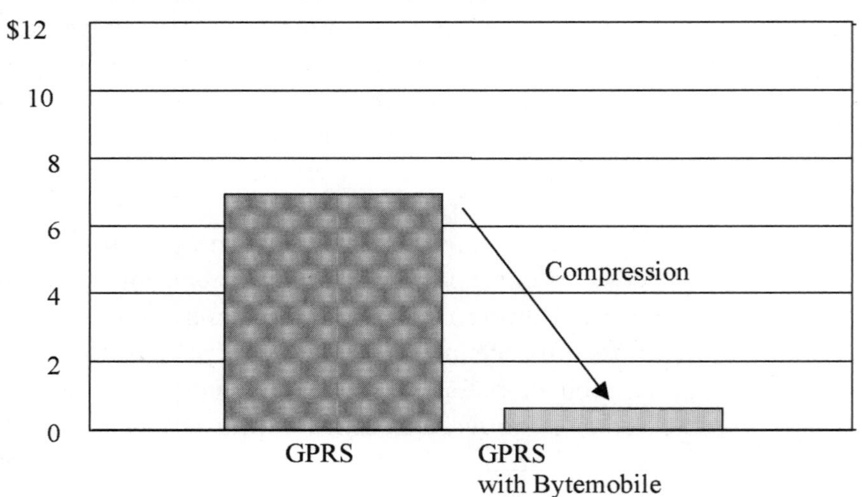

First, it is easier for content providers to access micropayments for their offerings in conjunction with an existing billing relationship such as what exists today between wireless service providers and subscribers. Second, the actual multimedia content will be delivered to the device. Finally, the type of content that is likely to appeal to audiences may in fact be targeted to very narrow audiences.

Other 3G applications could result in a bundled product. For example, Lobby7, a developer of public venue applications, has prototyped an application for a sports stadium where visitors could get instant replays, check player statistics, and order food and concessions. As such, the revenue stream for the venue would be expanded.

However, venue-specific applications are much more likely to use WLANs rather than WWANs due to their faster data rates and lower cost. Another company working on 3G-type applications is LightSurf Technologies, which takes digital photos with a camera the size of a square centimeter and then sends them anywhere in the world. The company built an infrastructure for Kodak and anticipates doing the same for carriers and handset manufacturers. Although the technology will work best on 3G networks, a good user experience is possible on 2.5G networks.

A Nokia 3G Application Case Study

How interested are the adult populations of eight countries in actually paying for 3G services? Table 2-8 illustrates Nokia's findings. The United States and Asia/Pacific (APAC) have the highest interest levels, driven more by those who are maybe interested. This is especially pronounced in the technologically hungry country of Taiwan, where over half of the adult population claims to be interested in purchasing 3G services. In the United States, the higher interest could be because they will likely be using the services described on the fixed Internet and are curious about receiving them on their mobile phone.

When you look at all the countries in terms of the most enthusiastic customers (very/fairly interested), there is a consistent group

Table 2-8

3G service interest by country

Country	Very Interested (%)	Fairly Interested (%)	Maybe Interested (%)
Global	6	8	24
The United States[4]	6	7	26
The United Kingdom	5	12	18
Belgium	5	10	19
Italy	5	10	13
Denmark	5	6	15
Europe (total)	5	9	16
Australia[4]	6	8	25
Singapore	5	10	23
Taiwan	6	11	36
APAC (total)	6	9	29

[4]The United States and Australia sample excludes 16 to 17 year olds.

Note: Total adult population sample (about 1,000 per country)

Source: Nokia

(5 to 6 percent) who are very interested at this stage and more variation in those who are fairly interested (from 12 percent in the United Kingdom to 6 percent in Denmark).

The following are some of the key factors that influence interest:

- **Internet usage (used the Internet for personal use within the last six months)** Those who are most familiar with the fixed Internet were more likely to be interested in paying for 3G services. The figures for Internet usage show the greatest difference in interest levels. Their overall interest level is 50 percent (combining very/fairly/maybe), whereas those who did not use the Internet have an overall interest of 29 percent. This difference in interest was much more apparent in the United States (47 percent against 24 percent) and Europe (45 percent against 22 percent) compared with APAC (57 percent against 42 percent). Overall, Taiwan showed the least differences with 64 percent current personal Internet users and 52 percent nonusers claiming to be interested in paying for 3G services.

- **Mobile phone usage** Existing owners are more likely to be interested in paying for 3G services (46 percent), but there is also significant interest from nonmobile phone owners (28 percent). The difference is more significant in APAC (55 percent against 30 percent) and Europe (40 percent against 19 percent). In the United States, there was not as much difference between the groups (42 percent owners and 36 percent nonowners). This is possibly because the lower mobile penetration levels mean that nonowners are more technically comfortable in the United States than in the other countries studied.
- **Gender** Fewer differences in interest levels were noted between the sexes (40 percent men and 36 percent women across the eight countries). Europe showed some larger differences (36 percent men and 25 percent women) and Denmark was the most pronounced (35 percent men and 17 percent women). Overall, though, it seems that the popularity of mobile phones with both sexes is likely to continue with 3G services.
- **Age** As may be expected, the younger generation is more interested in paying for 3G services. The extremely high interest levels come from the 16 to 22 year olds contrast to the lower excitement from those aged 55 and older. Youth interest is especially strong in the United Kingdom, the United States, Belgium, and Taiwan.

Who Might the Early Adopters Be?

To gain more insight into future uses of 3G services, the rest of the research study focused on the likely early adopters—mobile phone owners, Internet users, and those from ages 16 to 54. All of those interviewed in the remainder of the study are very, fairly, or maybe interested in purchasing 3G services. This group typically represents 10 percent of the adult population within the countries studied.

The early adopters spoken to within the adult population were much more likely to be interested in purchasing 3G services, have higher monthly mobile phone bills, and use the Internet more. This group was studied because they are more likely to have a better understanding of the types of services and their usage of 3G services. It was also decided to talk to those who pay or contribute to their mobile phone bill so that they were familiar with the existing tariff procedures.

Willingness to Pay for Specific Services

Which specific services are early adopters likely to pay for? E-mail is clearly the service that the early adopters were most willing to pay for (39 percent were very likely). Ticket booking, maps, banking, weather, Internet browsing, traffic information, and journey planning were all rated as likely to be purchased by many. Most of these are services they are likely to be familiar with (to some extent) at the moment and that they can rationalize as having a practical benefit. Many of the other services (especially video clip downloads, games, and gambling) had more of a niche appeal (for example, younger males) at this stage.

- **The United States** There was more enthusiasm for weather information (71 percent very/fairly) and journey planning (59 percent). Ticket booking (56 percent) and banking (45 percent) had less appeal than the global pattern.
- **The United Kingdom** Some differences appeared with Internet browsing (62 percent very/fairly). Shopping (50 percent) was higher, and weather information (45 percent) and maps (50 percent) were lower.
- **Belgium** Traffic information (66 percent very/fairly) and banking (57 percent) were more likely to be paid for overall. Weather information (44 percent), maps (50 percent), journey planning (40 percent), and shopping (35 percent) were greeted with slightly less enthusiasm.
- **Italy** More enthusiasm was shown for a number of services, such as e-mail (80 percent very/fairly), ticket booking (74 percent), Internet browsing (76 percent), maps (66 percent), traffic information (64 percent), and video downloads (37 percent).
- **Denmark** There was a higher likelihood than the overall for e-mail (84 percent very/fairly) and traffic information

(61 percent); it was lower for weather information (49 percent), maps (50 percent), and games (17 percent).

- **Australia** There was an increased interest for banking (66 percent very/fairly) and less interest in a number of other services, such as maps (53 percent), Internet browsing (48 percent), weather (48 percent), traffic information (48 percent), and journey planning (42 percent).
- **Singapore** There was a much higher likelihood of paying for gambling services (20 percent very/fairly) with some of the other services showing slightly less interest, such as e-mail (66 percent), weather (48 percent), maps (48 percent), music downloads (40 percent), and shopping (34 percent).
- **Taiwan** These results were most dissimilar to the global pattern. The Taiwanese were less likely to claim to be very likely to use services (although they were enthusiastic at the fairly likely level). Many services had higher interest levels, such as ticket booking (77 percent), maps (70 percent), journey planning (63 percent), banking (61 percent), music downloads (59 percent), and video clip downloads (42 percent). E-mail showed significantly less interest (57 percent), possibly due to the input of Chinese characters. Within the early adopters (and the confines of the sample size), there were some service bundle preferences beginning to emerge. One group (called *leisure junkies*) was very interested in games, music, and video downloads. Others were more keen on *mobile commerce* (m-commerce) or services that helped them while traveling. Those who represented the typical (more younger males) technology innovators (referred to as *boyz and toys*) were interested in most of the services, especially gambling and games. These findings are in line with previous Nokia 3G Research Centre results (see the article "Mobile Internet Study," November 2000, which is available at www.nokia.com/3G).

Tariff Preferences

How will consumers pay for these services? Flat fee and standard package structures emerge as the most preferred tariff structures, according to consumers. File size charging is the least preferred tariff structure (and is also the one that people are less likely to be familiar with). Research shows the high strength of preferences expressed with 45 percent remaining very unlikely to choose file size charging.

Respondents were not shown any price points for different tariffs. It may be that when these are considered, the standard package would be more popular than the (more expensive) flat fee. Overall, the order of preference was very similar for the different countries, although the strength of opinion varied. The Taiwanese were more enthusiastic about all of the tariff structures (especially subscription based).

Out of those interviewed, 9 out of 10 people said that their preferred tariff structure would encourage them to use 3G services. Adopting a tariff structure that the customer does not prefer may not stop them from using the services, but it will make it less likely that they will adopt the service.

Why a Flat Fee? When asked, respondents felt that the flat fee option was easy to understand, easy to monitor spending, there was no need to worry about time spent exploring, and you knew what was included. It represents the easy, no-risk option where you do not have to worry about the different costs.

Why Standard Package? The standard package enabled the respondents to know what was included, made it easy to monitor their spending, made it easy to understand, gave them more control, and enabled them to explore without worrying about cost. The benefits of this option do not seem to be very different from the flat fee, although there were questions about what would be included within the package.

Why Not File Size Charging? File size charging was seen as confusing and not something that customers identified with. They complained that they did not know the size of files, found it confusing, would not pay attention to file size anyway, and felt that they would lose track of their spending.

Spending Habits What would early adopters be willing to pay (for the 3G services mentioned) above and beyond their existing monthly mobile phone bill? Proportionally, the amount they were willing to spend represents an overall increase of above 50 percent (52 percent in the United States, 82 percent in Belgium, 67 percent in Italy, 64 percent in the United Kingdom, 70 percent in Denmark, 42 percent in Singapore, 36 percent in Australia, and 68 percent in Taiwan).

This increase in mobile phone spending is higher for 16 to 22 year olds ($27) compared to 23 to 34 year olds ($22) and 35 to 54 year olds ($22). Only a small difference in increased spending was found between the sexes ($24 for men and $23 for women).

Willingness to Pay for Premium Services

Two-thirds (66 percent) of the people interviewed said that they would be willing to pay an additional fee for using certain premium services (such as video clips, music to download, and information that was important to them). In Taiwan, the figure was 75 percent. This illustrates that many of the early adopters were not placing a fixed threshold on how much they would be willing to pay; if the content is there, then many of them claim they will pay for it.

Conclusions in the Nokia 3G Application Case Study

The following conclusions can be drawn from this case study:

- There is a strong interest among consumers in purchasing a wide variety of 3G services.
- Among early adopters (who pay their own bill), there is an apparent willingness to pay for a range of different 3G services.
- It is recognized that there will be a 3G hypermarket of services and the key for operator success will be in offering bundles that appeal to different segments.

- The early adopters claimed to be willing to spend a fairly substantial monthly sum on 3G services in addition to their existing mobile phone bill.
- A flat fee or standard package bundle was the preferred tariff structure. File size charging was the least popular option.
- Early adopter figures (for those who pay their own bill) indicate that they would be willing to pay an increase of over 50 percent a month if they receive the services they are expecting.
- Two-thirds of early adopters (who pay their own bill) would also be willing to pay extra for premium content of interest to them.

The Future of 3G

Europe and Asia are likely to be the areas that experience the first take-up of MMS. However, the real mass of subscribers will be achieved from 2003 onwards. It can also be stated that the Asia/Pacific market is likely to be strongly ahead of Europe in 2002 and will be led by Japan. Hence, Japan will also lead the way in MMS (although the region is already providing evidence of the issues involved in rolling out 3G services). Clearly, it also has the added advantage of already having a significant subscriber base running on PDC, which is a packet-based network on which MMS can begin to be rolled out.

Obsession with Technology

Despite claims to the contrary, the whole telecom industry suffers from an obsession with technology for the sake of technology. Some companies continue to exist on the basis of *jam tomorrow*—all their problems will be resolved by the next new technology. The telecom industry is neither market focused nor market led, and the mass adoption of SMS surprised the complete industry for exactly this reason. SMS only offers the ability to send a limited amount of plaintext. In a technology-obsessed environment, it could not possibly appeal to anyone. The market thought differently. The industry is now obsessed with 3G and MMS. The risk is that the industry has once again missed what the market wants now and will be taken by surprise.

Potential to Disappoint

Any eagerly awaited technology runs the risk of disappointing the market. This is especially true in the case of messaging where the users are able to experience a variety of formats to exchange messages. Any advance of messaging technology for mobile devices will be seen as an enhancement to SMS, but will be compared with e-mail and *Instant Messaging* (IM) experiences when the user compares feature sets. If the user is promised the ability to have pictures on their device, they might expect full-color, large-format *Joint Photographic Experts Group* (JPEG) images; however, if their device is only capable of 256 color 100×100 images, they will likely be disappointed and therefore feel negatively to the overall technology.

Next-generation subscriber forecasts will provide an indication of the take-up of MMS globally. MMS can, of course, be used on 2.5G networks; however, it is not until 3G networks that greater functionality will be seamlessly capable, such as streaming video and audio. As a result, the industry is likely to witness the take-up of MMS in a 2.5G environment—for example, a combination of simple photo messaging plus text for P2P messaging. However, the early stages will have limited functionality, and until users perceive that MMS provides sufficient additional value to upgrade their handsets, volumes will be low. This will limit the size of the MMSCs sold to very small systems.

Many industry participants believed that 3G networks across the world would not be fast enough to offer multimedia services when launched in 2002 (although this predominantly refers to services such as video or music clips).

The good news is that the next-generation wireless infrastructure being implemented supports the similar wired existing Internet solution. The bad news is that both wired and wireless products need to be further developed. In the end, the road to success does not lie in a person's ability to do everything right; they only have to do a few key things right.

SMS has been the pioneer leading the path in wireless data applications. SMS services were the first mobile data services that had a serious financial impact on the wireless service operator's revenue. The industry has seen tremendous growth and messaging will presumably continue to lead to profits in 3G (regarding wireless data) as well. The current split of total revenues from mobile applications will inevitably change and become much more versatile as new-/next-generation infrastructure is implemented and enables users to take full advantage of it.

CHAPTER 3

Wireless Messaging Today

This chapter discusses the wireless messaging formats, their characteristics and popularity, and why some formats are more successful than others—for example, i-mode (e-mail) in Japan and *Short Messaging Service* (SMS) in the Philippines and Europe. Additionally, this chapter examines how wireless messages are delivered to a device through the network and provides a comparison of the different devices used to message.

Some Important Attributes of Today's Wireless Messaging

Today's wireless technology is poised to fundamentally change the way people consume, interact, and organize their lives. The way consumers make shopping decisions, pay for cinema tickets, or conduct their daily business needs may never be the same again because of innovative message applications. This transition is depicted in Figure 3-1, which illustrates the replacement of wireline phone usage by wireless. One could say that wireless is pulling the rug out from under business practices geared toward wireline business.

When trying to gauge the current state of the wireless data market, users, analysts, and investors typically reference the current challenges of the *Wireless Application Protocol* (WAP) and point to Japanese wireless carriers (NTT DoCoMo in particular) as pioneers diagramming the course of opportunity. However, over the past few years, wireless messaging (namely SMS) has arguably had much more success in both Europe and Asia. It is important to note that the messaging craze in Japan is e-mail, not SMS. Japan has a fairly significant SMS user base, but NTT DoCoMo's high e-mail usage gives Japan its messaging notoriety.

In North America and Japan, several wireless carriers have implemented a competing messaging format to SMS—wireless e-mail. From a user's perspective, wireless e-mail appears similar to SMS, but there are two differences:

- First, an e-mail user must connect with the Internet, whereas an SMS user can send messages over voice networks through a

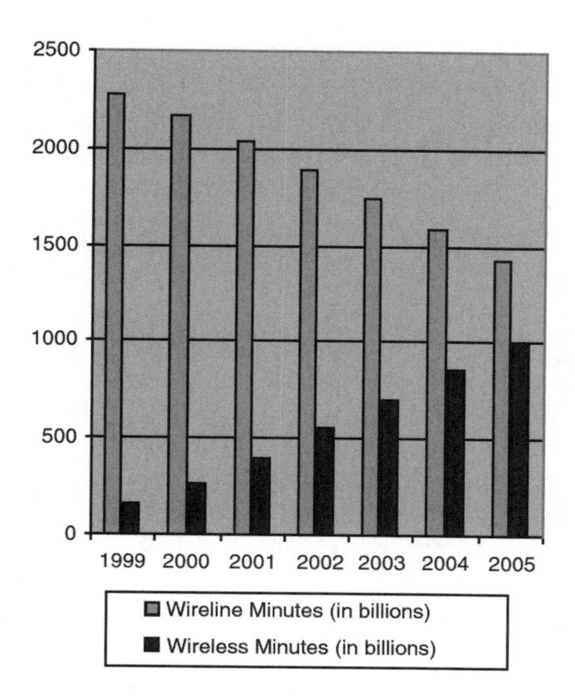

Figure 3-1 Wireless displacement of wireline phones minutes of use (Source: The Yankee Group, 2000)

control channel, which is parallel to the voice traffic channel. A connection to the Internet in wireless is typically made through a WAP session. As a result, e-mail will likely take longer to receive because it must connect to public *Internet Protocol* (IP) networks.

- Second, an e-mail user must type in an e-mail address, whereas an SMS user only has to type in the recipient's phone number.

Unlike in Europe, SMS messaging has experienced relatively modest penetration and growth in North and South America due to the lack of interoperability, although this has been recently amended. SMS and other forms of text-based messaging such as wireless e-mail are predicted to be the most heavily used wireless data applications. Table 3-1 lists the five main wireless messaging formats and their associated characteristics.

Table 3-1 The key characteristics of the five main wireless messaging formats available today

Type	Characteristics	Is Content Reformatting for Mobile Necessary?	Applications	Support	Timeframe for Availability
Short Message Service (SMS)	100–200 characters (dependent upon carrier and technology)	Yes	Simple *person-to-person* (P2P) messaging	Majority of phones	1990s
Enhanced Messaging Service (EMS)	Text messages plus sound, animation, picture, and text-formatting enhancements	Yes	P2P messaging with simple visual graphics	EMS standards expected to be widely adopted	2001
Multimedia Messaging Service (MMS)	Messages in multiple rich media formats such as video and audio plus text	No	P2P messaging with visual feel	MMS standards expected to be widely adopted	2002
E-mail	Typically 1,000 characters and can attach all file formats to transfer	Yes	P2P, and store and forward	All data-capable devices	1990s
Instant Messaging (IM)	Typically 20–30 characters	No	P2P or machine, and requires presence	WAP-enabled services	2001

Source: In-Stat/MDR

Regionality: The Regional Nature of Wireless

A wireless service, by the nature of the infrastructure on which it is built and by the control from the population of people that it serves, is tightly linked to its local market. Such local markets can, and do, determine the success of a wireless service offering in a marketplace. *Regionality* is a word used to describe this linkage of wireless services to their markets.

Regionality is an important quality to understand because it can help explain why a service might be successful in one place, but fizzle in another place. Knowledge of what makes a market place unique—especially when it comes to social or technical factors—has a direct impact on the adoption and success of a product. To create a better perspective about the wireless service offerings in markets around the world, it is important to understand the regional uniqueness and history of the market being examined. Examining the regionality factors can predict the success or failure of a product offering.

A great description of regionality can be found by examining Japan. In Tokyo, it is still fairly expensive to get a dial-up Internet connection even though the service is quite developed in terms of content. Out of this disparity comes the foundation for the phenomenal adoption rate of NTT DoCoMo's i-mode wireless data service. DoCoMo's i-mode burst a bubble of sorts. All of a sudden, Internet access was available in Japan. Before i-mode, an expensive dial-up connection was required to access the Internet—all that is required now is a cell phone. DoCoMo gave people a reason to purchase a cellular phone and data service; they filled a void. DoCoMo's success will be discussed in more depth later in the section "Japan's NTT DoCoMo i-mode Service."

The most advanced wireless data markets in Europe are in the Nordic countries, reflecting the advanced levels of wireless penetration and the willingness to be innovative and use that innovation. Furthermore, less expensive data versus voice correspondence and the right pricing of data meets the disposable income of users. Like Japan, Northern Europe has a high mobile phone market penetration and a fairly homogeneous network infrastructure. Additionally, the U.K. market also offers substantial upside potential, with business data applications generating a relatively high proportion of total revenues. The United Kingdom currently has the highest proportion of *business-to-business* (B2B)-enabled firms in Europe and hence has the potential to become a major wireless data market.

This indicates that there are some truly unique attributes in each wireless messaging marketplace around the world. Further details of these regional attributes are discussed throughout this chapter.

What Is Short Message Service (SMS)?

SMS is the most popular wireless data service in the world.

Brief History of SMS

SMS has the capability to send and receive basic text messages between wireless handsets via the control channel on a mobile network, which is parallel to the voice traffic channel. Each SMS message is limited to 160 Latin alphabet characters in length and drops to approximately 70 characters in length when non-Latin alphabets such as Arabic and Chinese are used. SMS was first incorporated into the *Global System for Mobile* (GSM) digital mobile phone standard. Transmission of the first SMS was sent in December 1992 from a *personal computer* (PC) to a mobile device via the Vodafone GSM network in the United Kingdom. In the United States, where the primary standards have historically been *Time Division Multiple Access* (TDMA) and *Code Division Multiple Access* (CDMA), two-way SMS was not introduced until late 2000. One-way messaging was available on TDMA networks since 1996, but it was used minimally.

Driven by the Consumer

SMS exploded in Europe even though service providers did not actively promote the service. Usage was driven by word of mouth, particularly with teens using prepaid service contracts. SMS was so successful early on in the prepaid teen market because the associated costs were approximately 25 percent less than a one-minute voice call. A simple example of an SMS would be "will be late" or "call when u land." In addition, ring tones and cartoons can be sent as SMS messages. Ring tones have been quite popular because they personalize the phone. Part of the value proposition of SMS is that messages can be sent from person to person, person to group, person to service, or service to person easily and quickly. For example, a subscriber could request any news on Tiger Woods' golf scores, and this information would be delivered via SMS with minimal delay.

Furthermore, low wireline Internet penetration is fueling SMS adoption. As a result, SMS is typically European youths' first messaging experience. In addition to P2P correspondence, SMS technology has grown to provide a remarkable range of services and information that can reach customers in new efficient ways. Enterprises have begun to use the service as a means to generate revenue. Dunkin' Donuts uses SMS in Italy to advertise donut specials. A user can save the coupon and use it for a discount the next time they visit Dunkin' Donuts. MobilKom Austria and the country's national rail service have successfully facilitated *mobile commerce* (m-commerce) through a partnership permitting passengers to purchase tickets via SMS while traveling on the train, saving them the time and hassle of waiting in line.

Ring tones are bringing further success for SMS. Nokia created a smart messaging protocol that was built on binary SMS rather than standard text SMS. Nokia had expected this technology to be used for information services and over-the-air service profiling. It languished for years, until it found its application in 2000—ring tones that enable users to change the way their mobile phone rings. Because the network operators were woefully inadequate and unable to offer the ring tone suppliers fair and flexible revenue sharing, the service providers started using premium-rate *interactive voice response* (IVR) voice platforms to trigger the transmission of ring tones. The ring tones market soon became a billion-dollar market.

The success of SMS was led by consumers, operators. This highlights an important fact for operators: When planning new offerings, they should not focus on dictating new services, but should focus on providing an open platform to the application's community and users, and let the market determine success. This will be the easiest way to build scale and develop a network effect for the service. SMS has prospered due to a number of factors:

- **Adoption by the youth segment** The youth market is an early adopter and helps build scale and mass acceptance of new services.
- **Prepaid services** Prepaid lowers the cost of entry and helps build scale for SMS services. SMS would still be a success without prepaid services, but it would have taken longer to gain mass adoption.
- **Affordable service** The cost of messaging is significantly cheaper than per-minute mobile calls, thus encouraging high-volume usage.
- **Common GSM platform** The uniformity of one network technology allowed for seamless message correspondence, unlike in the United States where competing technologies have only complicated SMS messaging development until recent interoperability changes.
- **Interoperability** Interoperability has been a proven driver of messaging growth. In the United Kingdom, once interoperability was available in April 1999, SMS traffic increased sevenfold.
- ***Calling Party Pays* (CPP)** CPP encourages the use of messaging as users do not have to pay for incoming messages. In the United States, two-way billing has slowed messaging and overall wireless growth as the receiver has to pay for receiving messages. CPP does not translate to lower revenues since incoming messages stimulate response and more traffic.
- **Wide diversity of services** With a common platform and large subscriber base, a large variety of services have been developed, the most popular of which are consumer applications such as
  - Simple P2P messaging
  - Notifications of voice and fax mail
  - Ring tones

- Information services (weather and news)
- Mobile banking and payment
- National and international roaming

SMS Pricing

SMS pricing has been revolutionary in Europe and is now available in Asia. It is the first mobile telephony service not to be priced according to the time of day, distance, or length of time. It follows a pay-per-use model for *mobile-originated* (MO) messages. An identical rate is applied for all messages, whether they are 1 or 160 characters. SMS is currently responsible for approximately 10 percent of total revenues across all European carriers. MO messages require a European average of $0.10 per message, whereas *mobile-terminated* (MT) are free, unlike in the United States where the user is charged to send and receive messages. SMS messages also may prompt a voice response, which creates further revenue for providers. Currently, in Europe and Asia, no operator-to-operator charges are levied for SMS interconnection. This is based on the assumption that roughly similar volumes of SMS traffic are exchanged between networks. This may change as traffic volumes begin to vary and some carriers demand to be compensated for carrying excessive amounts of competitors' SMS traffic.

Currently, bucket plans are the most commonly found form of SMS billing in the Americas, although some carriers have introduced per-message charges and unlimited bucket message plans. In-Stat/MDR estimates the average cost of an SMS in the Americas is $0.05 and expects that price to decline to approximately $0.03 in 2005. Although the majority of U.S. mobile operators initiated their SMS offering through bucket plans, the European and Asian pay-per-use tariff strategy is beginning to surface because SMS user levels are significantly below carriers' estimates. For example, Cingular Wireless offers four different SMS pricing plans. The first is the European model—CPP. The user is charged $0.10 per MO or MT SMS, although there is no required monthly subscription fee. The second, third, and fourth plans offer 100, 250, and 500 MO or MT messages for $3, $6, and $10, respectively. Each message exceeding the allotted amount costs $0.10.

The average European MO SMS fee is approximately $0.10, with no fee for MT SMS messages—a CPP model. The majority of the Asian providers have adopted the European SMS tariff strategy primarily due to the European influence (investments) in the countries' carriers. Providers in the Philippines offer a set number of MO SMS for free each month (approximately 500); excess messages cost approximately $0.01 per message. This strategy has led to record SMS volumes exceeding 37 million per day and 32 percent of the Globe's revenue. SMS usage has become so popular that traditional wireline phone providers have implemented wireline SMS capabilities. Japanese carriers, who already have a successful messaging base through e-mail, require a monthly fee of 100 yen and charge 0.3 yen per packet for SMS. One SMS message typically requires three packets, which adds up to 0.9 yen. Table 3-2 illustrates In-Stat/MDR's SMS MO pricing forecast.

Providers in the Philippines, hoping to generate more revenue from SMS, have discussed decreasing the amount of free SMS mes-

Table 3-2

SMS per message pricing forecast

	Monthly Service Fee	2001	2002	2003	2004	2005
Americas*	NA	$0.05	$0.04	$0.04	$0.03	$0.03
Europe	NA	$0.13	$0.11	$0.09	$0.08	$0.07
Japan (per-packet message)**	$0.80	$0.0072	$0.0072	$0.0072	$0.0072	$0.0072
Rest of world	NA	$0.02	$0.02	$0.02	$0.02	$0.02

Source: In-Stat/MDR

*Calculated per-message fee by average allotted monthly messages allowed divided by the monthly fee

**Assuming three packets equal one message

sages per month, which has been met with tremendous opposition from subscribers. In China, Korea, Australia, and South Africa, MO SMS is beginning to be adopted. This can be attributed to the following factors:

- SMS phones are becoming more affordable.
- Subscribers are learning about the service.
- MO SMS is less expensive than an outbound voice call.
- It is easy to register for SMS services.

The price of SMS in the Philippines, Singapore, China, and Korea is attractive, although providers in Thailand, Taiwan, and Indonesia could do more to promote the service, such as permitting interoperability between carriers.

Early Mistakes

Early mistakes include the technological loophole in SMS and the current lack of prepaid SMS roaming. In this day and age, with the Internet revolution spreading information, it is certain that people will identify loopholes in technology and expose them to their advantage. They did it with SMS in wireless technology. Suddenly, millions more SMS messages were being sent with individual mobile phone subscriptions accounting for thousands of SMS per month alone as they set up automated message generators. Network operators worked with their platform suppliers to try to sort this out and charge prepay customers for SMS. Meanwhile, SMS incubated and spread, and people began using it because it cost nothing, whereas carrying out the same transaction using voice had a price. Eventually, after a few months, the network operators finally got their act together and managed to charge prepay users for SMS so they could decrement the prepay credit by the cost of an SMS message.

A mass SMS message distribution campaign was then typically sent out such that people who had used SMS received a text message informing them that from a certain date, they would be charged for SMS. This led to an immediate and protracted decline in SMS usage to between 25 and 40 percent of the precharging levels as people suddenly stopped using SMS or using it as much. Then something interesting happened—the volume of SMS messages started gradually increasing again and soon reached its precharging levels. SMS volume growth has continued its upward growth ever since, fueled by simple P2P messaging as people told each other how they were feeling and what they were doing—information services and other operator-led initiatives failed to interest the user community to any degree. Although it was free, SMS had become an important part of the way that people communicated with each other in their daily life. SMS would have taken off without this prepay factor because it was already being used before that time, but it would never have taken off as quickly.

The introduction of prepay mobile tariffs in which people could pay for their airtime in advance and thereby control their mobile phone expenditure was the catalyst that accelerated the success of SMS, particularly because the youth market adopted prepaid service plans and could correspond less expensively via SMS than voice. The mobile operators were unable technically to bill prepay customers for the SMS they were using because the links between the prepay platform and the billing system and the *SMS Centers* (SMSCs) were not in place. Today, prepaid SMS roaming is still an inhibitor to greater usage and revenue from prepaid users. The mobile operators responded with silence—the prepay literature did not mention SMS at all even though the prepay phones supported the service.

Components of an SMS Message

The components needed to send an SMS message include

- A PC or cellular phone with the correct application
- An SMSC on the hosting carrier with the service enabled

The Path of an SMS Text Message

The path of a SMS message depends upon the sender's device. If the sender's device is a wireline PC, the SMS travels via the Internet to the appropriate carrier's Internet gateway, which is connected to the sender's carrier's SMSC. If the sender's device is a mobile handset, the message travels back to the sender's carrier's SMSC and is distributed to the receiver's handset if they're using the same carrier for service. If the carriers are different, the SMS is sent from the sender's SMSC via the Internet to the receiver's SMSC and is then distributed to the recipient.

What Is an SMSC?

In order for a wireless carrier to offer SMS and unite the messaging path between mobile phones, the carrier needs to purchase an SMSC from the wireless infrastructure suppliers CMG, Logica, Comverse, or Openwave. The SMSC is a central store-and-forward facility that accepts, buffers, processes, and distributes all SMS messages in a mobile network. All SMS messages travel through the SMSC. Figure 3-2 illustrates how the SMSC is linked together with the

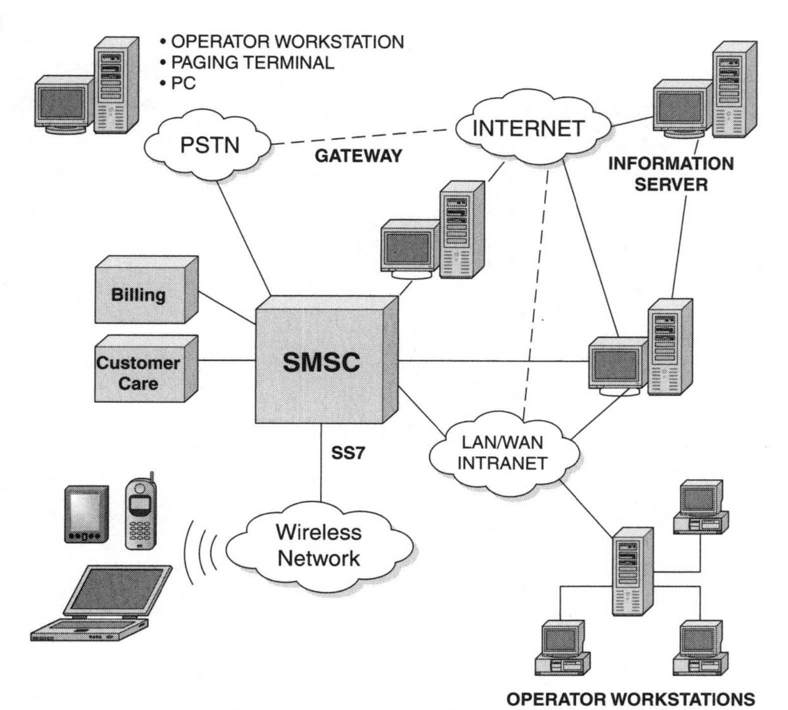

Figure 3-2 SMSC interconnections with other network elements (Source: In-Stat/MDR)

Internet and a company's intranet, wireless network, and billing systems. SMS is a store-and-forward technology. If the recipient's wireless device is dormant when the SMSC attempts to push an SMS, the SMS message will be stored within the SMSC for approximately 72 hours before it is deleted. If the recipient's device is turned on within 72 hours, the SMS will be sent automatically.

SMS also features the confirmation of message delivery to the SMSC, but not the receiver, which is unlike paging, where a sender is guaranteed transmission of the message and will receive a confirmation once the message has been received.

How to Send a Text Message

To send an SMS message, recognizing that all wireless devices differ, a user typically scrolls to select a menu option from a list, selects messaging, inserts the phone number and then the message, or vise versa, and hits send. The message is sent. Figure 3-3 provides a basic example of how this is done.

The youth market has adopted a form of code that makes correspondence quicker. A new alphabet has emerged because it took a long time to enter full sentences. Abbreviations such as "C U L8er" for "See you later" sprung up to save time and portray a cool persona. The use of smiles to reduce the abruptness of the medium and help indicate the mood of the person in a way that was difficult with just text became popular. Table 3-3 provides shortcuts or abbreviations that are used to speed up typing and exhibit some emotions.

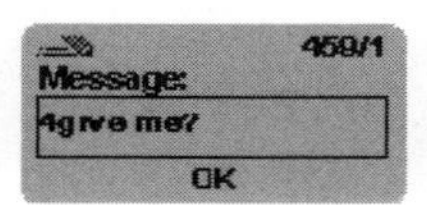

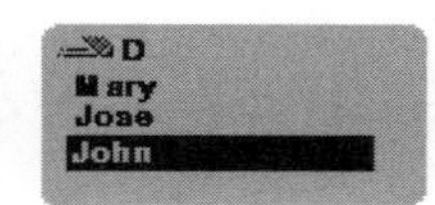

Figure 3-3 Cell phone screen shots (Source: In-Stat/MDR)

Table 3-3

Text examples

<table>
<tr><th>Original</th><th>SMS version</th><th>Original</th><th>SMS version</th></tr>
<tr><td>Anything</td><td>NTHING</td><td>Are you OK</td><td>RUOK?</td></tr>
<tr><td>Be</td><td>B</td><td>Before</td><td>B4</td></tr>
<tr><td>Be seeing you</td><td>BCNU</td><td>Cutie</td><td>QT</td></tr>
<tr><td>Date</td><td>D8</td><td>Dinner</td><td>DNR</td></tr>
<tr><td>Easy</td><td>EZ</td><td>Eh?</td><td>A?</td></tr>
<tr><td>Excellent</td><td>XLNT</td><td>Fate</td><td>F8</td></tr>
<tr><td>For</td><td>4</td><td>For your information</td><td>FYI</td></tr>
<tr><td>Great</td><td>GR8</td><td>Late</td><td>L8</td></tr>
<tr><td>Later</td><td>L8R</td><td>Love</td><td>LUV</td></tr>
<tr><td>Mate</td><td>M8</td><td>Please</td><td>PLS</td></tr>
<tr><td>Please call me</td><td>PCM</td><td>See/sea</td><td>C</td></tr>
<tr><td>Rate</td><td>R8</td><td>Speak</td><td>SPK</td></tr>
<tr><td>See you later</td><td>CU L8R</td><td>Thanks</td><td>THX</td></tr>
<tr><td>Thank you</td><td>THNQ</td><td>Today</td><td>2DAY</td></tr>
<tr><td>Tomorrow</td><td>2MORO</td><td>Want to</td><td>WAN2</td></tr>
<tr><td>What</td><td>WOT</td><td>Work</td><td>WRK</td></tr>
<tr><td>Why</td><td>Y</td><td>You</td><td>U</td></tr>
<tr><th colspan="4">Emotions</th></tr>
<tr><td>:-)</td><td>Happy/smiley</td><td>:-||</td><td>Angry</td></tr>
<tr><td>:--))</td><td>Very happy</td><td>%-)</td><td>Confused</td></tr>
<tr><td>;-)</td><td>Winking</td><td>:-&</td><td>Tongue tied</td></tr>
<tr><td>:--(</td><td>Sad</td><td>O:-)</td><td>Saintly</td></tr>
<tr><td>:--D</td><td>Laughing</td><td>:'-(</td><td>Crying</td></tr>
<tr><td>:--O</td><td>Surprised/shocked</td><td>:-@</td><td>Screaming</td></tr>
<tr><td>:--*</td><td>Kiss</td><td>:-P</td><td>Tongue in cheek</td></tr>
<tr><td>:-@)</td><td>Pig</td><td>*:-)</td><td>Clown</td></tr>
</table>

Source: In-Stat/MDR

GSM Versus Non-GSM SMS Architectures

How have 18 individual European countries with just as many distinct languages had such a high SMS usage? The answer is one technology—GSM. Initially, European SMS could not be transmitted between two subscribers of different networks. European providers formed interconnection agreements with their European counterparts to enable the free movement of text messages between most European networks. Once SMS messages could be exchanged with other GSM networks, SMS began to surge. Figure 3-4 illustrates the interoperable effect on SMS in the United Kingdom.

European carriers have a distinct advantage over their North American counterparts—the uniformity of Europe's GSM network. U.S. carriers must buy spectrum licenses from the *Federal Communications Commission* (FCC) to construct their network in specific parts of the country. With GSM, licenses are not as important because messages are exchanged freely and adequate coverage is available, making every GSM carrier in Europe appear as though it has its own Europe-wide network. Additionally, North American carriers have not agreed on SMS roaming as with voice, which presents another European advantage. Lack of SMS roaming, for example, would leave a Verizon subscriber without SMS capabilities in an area only covered by Sprint, AT&T, or an analog carrier. Operators in

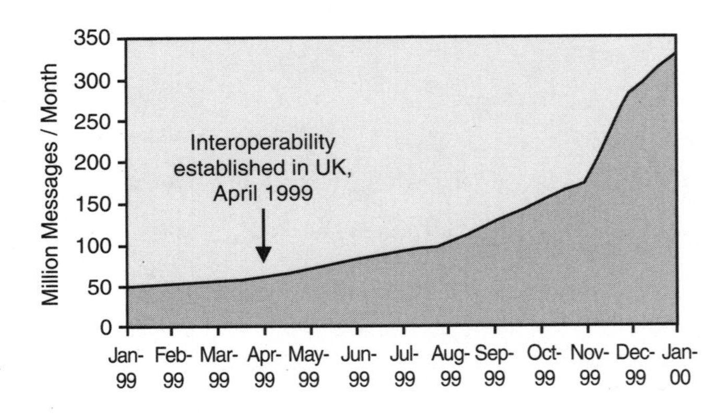

Figure 3-4 This figure illustrates an important fact—that SMS market up take is tied to interoperability between wireless carriers. (Source: EMC)

the United States are not as concerned with SMS roaming because each one desires to set up a nationwide network.

Although GSM conflicts with CDMA and TDMA technologies, it is technologically feasible to provision the networks to exchange SMS messages. Network technology is not the limiting factor in offering SMS on the same network; the interoperability of the different network technologies causes a problem. Typically, carriers wanting to transmit messages across CDMA and GSM must employ a secondary server to translate the messages for the other network technology. Companies such as Infomatch and CMG offer these services.

In addition to having different technologies, capacity constraints exist. Capacity is more of an issue in the United States than it is in Asia or Europe. Asia and Europe approach capacity issues when SMS is used in conjunction with mass marketing campaigns through either the printed press or TV. Operators report tremendous surges when an SMS number is displayed on the TV to win a free prize, often paralyzing their SMSC for a brief period. In the United States, operators constantly struggle with capacity and often kick off users. Many carriers are struggling to provision adequate capacity to handle their top priority—voice—in addition to SMS and e-mail. Additional SMS traffic, plus competitors' SMS traffic, may be too much data traffic to handle for a network designed for voice. A more in-depth discussion of GSM and non-GSM architectures appears in Chapter 7, "Wireless Messaging Infrastructure."

The main point to take away from the common architecture setup in Europe is that it enables carriers to move faster in bringing to market innovative messaging applications that can further drive usage and revenue. Messages do not need to be manipulated in order to send them to a different network technology. Two examples of this include EMS and MMS.

The Coming of EMS

The next step in the evolution of SMS is EMS (sometimes known as SMS+), which is capable of sending a combination of simple pictures and animations, sounds, and text as a message that can be displayed on an EMS-capable device. The accident of SMS has opened the doors for EMS and MMS. The craze of SMS has led the majority of wireless industry observers to believe that more advanced formats of

SMS such as EMS or MMS will achieve similar success. EMS delivery path is identical to SMS with the store-and-forward technology through the SMSC and the signaling channel. As the replacement of SMS- for EMS-capable devices begins, an intelligent converter will be required to reformat EMS messages for SMS-capable devices. EMS will be discussed in more depth in Chapter 8, "Wireless Messaging's Future: A Look Down the Road."

The Next Phase—MMS An enhancement to EMS is MMS, which incorporates richer capabilities such as audio and video in messages. A user can combine text, pictures, photos, animations, speech, and audio for the ultimate messaging experience. Furthermore, a user can take a digital picture and customize it with a favorite song and fonts.

MMS builds on the successful message-push paradigm of SMS and enhances communication possibilities for mobile users. MMS has the capability to receive, edit, and send images. This empowers users in all areas of life, enhancing personal connectivity. Images, sound, and video sequences can also be stored in the phone for later use. Wherever the user is and whatever the user is doing, MMS brings the power and freedom of complete communication to one's fingertips.

Instant Messaging (IM): What Is It and Where Is It Going?

As technology evolves, the wireless Internet will permit greater functionality. The characteristics of the wireline Internet will be found wirelessly. One application that has found success in wireline technology and will find success in a wireless world is IM. IM extends the attractive aspects of SMS as a communication tool. It eliminates the majority of the network delay created by Internet routing bottlenecks while providing one's peers with presence capabilities. Currently, wireline IM is based fundamentally on a closed system where a user can only correspond with another IM user utilizing the same network (for example, an AOL end user to AOL end user).

The end user's ability to see if another end user is available for correspondence (presence) differentiates IM from SMS. All SMS users know that a voice call is more instantaneous than SMSing someone repeatedly. If an SMS user knows that the person whom he or she is trying to contact is present and available, a quick voice call will be more productive than several messages, which often can exceed the price of a one-minute voice call and require more effort.

A typical user interface to access wireless IM requires the end user to have a WAP-capable device, although IM applications are now being sent through an SMSC as SMS messages. This allows for a different billing format—per-message versus duration corresponding. Through WAP, the user initiates a WAP session and logs into the appropriate IM service. The experience is quite similar to logging onto the Internet with a dial-up modem and an IM service provider, except the computer hard drive and monitor is, in comparison, a tiny phone with a slower processor and significantly smaller screen. In a *General Packet Radio Services* (GPRS) always-on, always-connected network service, for example, the presence list of subscribers will be one or two screens from the home screen and constantly available. Presence, a significant component to IM, has the capability to alter how society corresponds. People attempting to correspond with another person will have the option of contacting that person directly or leaving a voice or e-mail rather than having to track the person down. Figure 3-5 provides some examples of a wireless IM session.

Before IM comes to the wireless world, its providers need to ensure that wired tactics are not repeated in the wireless one. IM cannot become a widely used application until it clears one hurdle: interoperability. In the wired world, IM providers have ensured that users of their services cannot communicate with users of their competitors' services. Both the established IM providers and the carriers need to ensure that this hurdle is overcome in the wireless world for the good of the application and wireless development as a whole. U.S. wireless phone users have already had to sacrifice the ability to correspond via SMS and carriers have missed out on the revenue

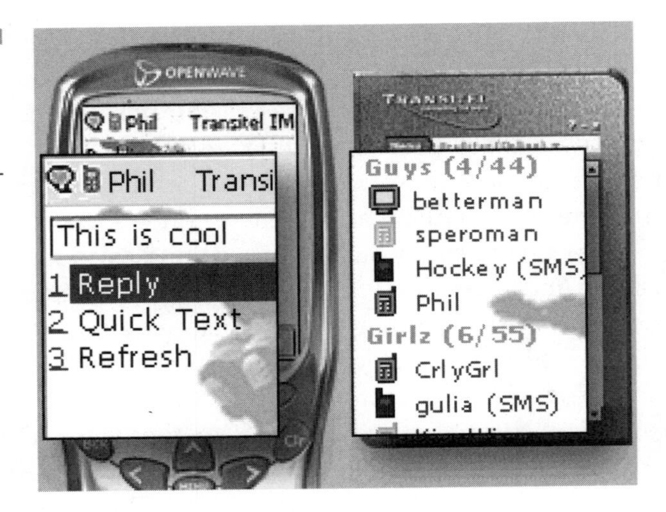

Figure 3-5 Wireless IM screen shots (Source: Openwave)

opportunity. The industry should make sure this mistake is not duplicated.

IM and Presence Interoperability

Currently, presence interoperability does not exist between the most popular wireline IM providers, although it is being discussed. This remains the case in a wireless world. If carriers are going to create an appealing new messaging service and a new revenue stream through IM, they must extend the value of their network infrastructure through interoperability. Wireless IM can be that service, although presence is the key component to its success (assuming interoperability is addressed). Carriers hope to alleviate the presence problem through several different protocols such as the *Profile for Instant Messaging* (PIM), *Presence and Instant Messaging Protocol* (PRIM), or *Session Initiation Protocol* (SIP). A sever-to-server protocol is a more practical solution for interoperability than a client-to-server protocol. The client-to-server protocol cannot be easily replaced. To date, SIP has been the most talked about answer to the presence interoperability issue.

A number of industry Working Groups have developed components for the improvement of wireless IM including Wireless Village and the PAM Forum. The Wireless Village (their initiative is called *Instant Message & Presence Services* [IMPS]) was formed in April 2001 to help create a common protocol and promote interoperability between wireless and wireline IM users. Through its founders, Ericsson, Motorola, and Nokia, the Wireless Village initiative aims to create an environment where wireless and wireline messages integrate seamlessly. This initiative, containing the major market share of active handsets worldwide, could have a significant impact on the development of IM.

The PAM Forum consortium has members from the voice, data, and wireless networking services and applications community. The PAM Forum's primary purpose is to create the *application programming interfaces* (APIs) associated with wireless IM. The PAM Forum's APIs and the Wireless Village's protocols will complement each other's work.

The Value Proposition of IM

Mobile devices and services are decreasing in price, and *quality of service* (QoS) and reliability have improved compared to a few years ago. If mobile device service, pricing, quality, and reliability are below wireline pricing, quality, and reliability, why has the industry experienced such rapid growth? Because of the value of wireless mobility. Mobile devices can be reached while not in one specific place. Subscribers are willing to sacrifice premium service for mobility. The value and luxury placed upon a wireless solution, combined with all the flaws, outweighs the superior service and pricing of wireline technology. Users are willing to tolerate sub-par service for mobility.

Now that the value proposition for wireless has been presented, the value proposition for IM is unique to each user. The majority of IM users find value in knowing who is and is not available for correspondence, and if they are available, users appreciate the ease and speed IM provides. Numerous wireless analysts estimate that the presence feature of IM would at minimum double the number of total messages sent by a user. In addition to the presence feature, the capability for messages to intelligently find and follow someone based upon their set preferences is also an attractive development.

Find me, follow me technology enables an end user, based upon certain permissions, to select the device, communication state, times willing to receive correspondence, and the preferred method of correspondence. Physical location could refer to a wireline or wireless situation. Communication state refers to whether the end user is currently engaged in correspondence or free. The willingness to receive correspondence refers to whether the end user desires to be contacted, which is also provided through permissions. The preferred method of correspondence refers to whether it is preferable to contact the end user with voice or text and at home or on a mobile device. Find me, follow me services will only be successful based upon the level of detail and ease of configuring one's personal profile. These are historically the reasons why find me, follow me services have not yet been successful. This scenario invites correspondence through presence. Voice and e-mail messages could decrease significantly and productivity could increase many times just by knowing the state of one another.

IM Alliances: Is the IM Log-Jam Breaking Up?

The most well-recognized wireline IM providers have signed partnerships with wireless carriers and pay them money in exchange for provisioning their IM services wirelessly. This extension is the necessary next step in the evolution of IM. The fundamentals of IM perfectly suit a wireless environment with its text-based media and its immediacy.

The majority of wireline/network provider IM partnerships (for example, AT&T and AOL) are primarily a marketing campaign for the network providers and will be short lived until greater revenue opportunities arise. Wireline IM providers and wireless network providers possess identical motives for signing these deals. Each company desires to be the first one to market by appearing more sophisticated than their competitors and to create subscriber growth while limiting churn through new *value-added services* (VASs). These partnerships have so far been successful in increasing subscribers, usage, value, and ultimately revenue.

In addition to increasing subscribers, usage, value, and revenue, these partnerships are a good training ground for more enhanced mobile Internet applications and services for the carriers. Once carriers have achieved a certain level of competence with wireless IM (wireless data) through the already established wireline IM partnerships, additional IM partnerships between carriers and IM specific vendors will be created—for example, Followap or MessageVine.

Wireless carriers may partner with IM vendors to launch their own interoperable service instead of continuing to promote their already-established wireline IM service provider partnerships for the following reasons:

- *Wireless carriers receive a larger share of the revenue.* By offering their own brands of IM services, providers can gain increased revenue through IM, SMS, and e-mail usage.
- *Carriers increase their brand value and reduced churn.* Carriers who have partnered with traditional wireline IM service providers may find that their brand competes with their partner's. This would effectively eliminate any benefits the carrier might have gained from wireless IM, they have greater control over service development, improved security and reliability, and they could leverage complementary services such as *location-based services* (LBS) and m-commerce.
- *Carriers are usually the owners of subscriber location, profiles, presence information, and billing relationships (which is possibly the most important relationship).* All of these can be leveraged to create a compelling service for end users. IM can add value to carriers and end users through an array of new services facilitated by presence. Location, m-commerce, and advertising are some examples. It is up to the providers of these services to ensure that they add value to the consumer.

So far, nothing has changed in the wireless IM world. Interoperability is still not mainstream. In fact, IM has taken a bit of a back seat to SMS as carriers seek to squeeze all they can out of SMS by now trying to add more content, such as advertising. IM should become more of a main feature when next-generation devices and networks couple always-on capabilities with attractive pricing.

WAP: The Mobile Internet

WAP is a standard or protocol for wireless devices and the accompanying infrastructure equipment. WAP provides a standard way of linking the Internet to mobile phones, *personal digital assistants* (PDAs), and pagers/messaging units. WAP is based on a client-server approach, making wireless devices the ultimate thin-client platform. A microbrowser is built into the wireless device and has limited system requirements. The majority of the processing resides on the WAP gateway and leverages the servers and other equipment of the wired Internet.

The WAP Forum was formed in December 1997 by Ericsson, Motorola, Nokia, and Unwired Planet. (Since that time, Unwired Planet changed its name to Phone.com and in November 2000, Phone.com merged with Software.com, resulting in the new company Openwave Systems, Inc.) The WAP Forum has become the industry association, comprising over 630 members, and has the following stated goals:

- To bring Internet content and advanced data services to wireless phones and other wireless terminals
- To create a global wireless protocol specification that works across all wireless network technologies
- To enable the creation of content and applications that scale across a wide range of wireless bearer networks and device types
- To embrace and extend existing standards and technology wherever possible and appropriate

The WAP Forum does not develop products; rather, it creates license-free standards for the wireless industry to use to develop products. The latest specification is published on the WAP Forum web site (www.wapforum.org).

WAP is based on *Wireless Markup Language* (WML) and *Handheld Device Markup Language* (HDML). WML is a subset of *eXtended Markup Language* (XML) developed by the *World Wide Web Consortium* (W3C). All WML traffic over the Internet uses the standard *Hypertext Transfer Protocol* (HTTP) 1.1 format when communicating between Internet nodes.

The WAP specification continues to evolve and improve. Version of WAP 1.2 incorporates push technology, enabling interactive services to be provided. Support for secure payments is provided by *Wireless Transport Layer Security* (WTLS), the WAP equivalent of *Secure Socket Layer* (SSL) protocols that are employed for e-commerce on the standard World Wide Web sites. WAP 2.0/WAP *next generation* (WAP NG) uses a standard known as *eXtended Hypertext Markup Language* (X-HTML), which has backwards compatibility to WML. The new standard will consist of the convergence of HTML, WAP, and i-mode (cHTML or Compact HTML) standards.

The WAP Forum has successfully started a camp of software and hardware developers focused on wireless data, something that had not achieved critical mass prior to WAP. Openwave, one of the founders and promoters of WAP, has over 160,000 registered developers in their Developer Program.

In addition to supplying software to network operators, Openwave supplies microbrowsers to mobile phone operators. By the end of December 2000, it had supplied products for 12.1 million mobile phones, which was up from 6.9 million at the end of September 2000. Infowave announced the latest editions of its browsers, version 5.0, which incorporate a *graphical user interface* (GUI), color screens, and pop-up menus. This will make the appearance of phones based on WAP 5.0 much more like a mini-PC screen based on Windows. Prior versions of WAP supported text menus that were black and white (much more like DOS on PCs in the early 1980s). The new browsers will make WAP services much easier for subscribers to navigate and use.

In February 2000, Ericsson announced formation of the Ericsson Developers' Zone Alliance Program to create business opportunities for the participants in four areas including WAP. In January 2001, Ericsson announced the setup of Ericsson Internet Applications and Solutions AB to develop service networks and application solutions.

Nokia has established the Nokia Active Developer Program, which focuses on supporting application or solution developers while developing mobile Internet or intranet applications, services, or solutions based on Nokia software and service platforms. More recently, Nokia announced a global cooperation agreement with Anderson Consulting, whose name has now changed to Accenture, designed to help corporations and service providers to build their own WAP-based wireless solutions. Nokia has also made agreements with Compaq Computer, Siebel Systems (a leading supplier of e-business applications software), SAS Institute (a leader in e-intelligence and data warehousing), and Germany's SAP AG (a vendor of B2B software solutions all aimed at developing and deploying mobile WAP solutions).

Although the standard is widely supported within the industry, development and adoption of WAP services has been slow, and deployment of WAP-enabled phones remains low. WAP and interactive wireless services are suffering from a case of overbilling—meaning there was a lot of hype and underdelivery. A technical description of WAP will be discussed in Chapter 7.

The Mobile Internet Hype Debacle

A shortage of compelling content and slow connection speeds has left a few early adopters disenchanted. With weak adoption and significant cost control pressures, content developers and carriers are scaling back their wireless Internet initiatives in favor of other opportunities with more attractive near-term revenue-generating possibilities, such as messaging. Communication applications are being used and generating revenue as opposed to Internet content, with the exception of DoCoMo. With few compelling services available, adoption will likely remain low. It's the old chicken-and-egg conundrum. Users will not appear until the services are better. With so few users, there's little incentive for carriers to develop more applications.

Two theories have evolved regarding application development. One camp develops applications based on the current technology. Like the Atari of video games, these services remain fairly basic. Meanwhile, the other camp remains focused on developing for technologies that may exist in the future, but aren't currently available. With a bevy of companies all attempting to capitalize on the wireless opportunity, those developing for now stand a greater chance of succeeding than those developing for tomorrow. With *third-generation*

(3G) rollout schedules that seem to get pushed back further and further, tomorrow is likely to turn into the day after and then the day after that. Very few of these companies will have the capital and staying power to hold out until the day that 3G and broadband become commercially available and widely deployed.

No matter what appears on a mobile screen, it needs to be small in size and short in length. Stock quotes, temperatures, and price comparisons are all examples of information that can be stated concisely. Currently, surfing the Internet on a WAP phone is highly impractical as there can be blocks between the gateway and web sites, causing delays. More fundamental are the form factor restrictions that allow only limited improvements. These restrictions are certainly true of WAP phones. Other mobile devices, such as Palm Pilots, could offer comparatively greater convenience of use. The frenetic activity being seen in marketing new devices should lead to enhanced levels of user-friendliness. Effective voice-recognition techniques could remove the need for many present-day keypad functions. Vodafone and other players are investigating the incorporation of these techniques into mobile devices.

Connection charges are still relatively high, dissuading enthusiastic users from using the Internet on a WAP device. Always-on connections should remedy this situation somewhat. The most effective applications and those for which a consumer will be willing to pay a premium are those where information is extremely time sensitive and where the benefit to the consumer of precise, timely information outweighs concerns about the cost of airtime (for example, online stock quotes). As the cost of handsets has steadily decreased, wireless penetration has grown rapidly and operators anticipate that this trend will continue by subsidizing handsets as the move to a data environment progresses.

Wireless Internet-based technology is one of the clearest examples of convergence currently in the marketplace. Convergence is the meeting of two distinct technologies (in this case, mobile phones and Internet browsers). As the communications power of mobile devices increased in step with processing power, it became viable to deliver digital material to mobile users via this platform.

The key is to see the WAP channel as a nonmedia channel. It is personal data access technology—a tool for enhancing business productivity and making your personal life that much easier. It is an immediate and targeted information source providing the vital data you need specifically for your location—from the current stock price of your employer to the screening times for *Charlie's Angels*. This technology will capture the hearts and minds of Generation Y and beyond.

SMS and WAP Linkages

WAP and SMS are not competing technologies. Rather, WAP and SMS complement each other. For instance, SMS does not run via the Internet, whereas WAP was created to help handsets interact with the Internet. Thus, a user can send a message over a standard wireless phone connection using SMS and then use a WAP browser to get a stock quote from a web site; however, as technology improves, users will be able to use SMS for a stock quote. Similarly, wireless e-mail is sent by establishing an Internet connection, whereas SMS does not need an Internet connection because SMS runs over standard wireless voice networks. Regardless of whether a text-based message is sent in the form of SMS or e-mail, the wireless messaging market will continue to experience heavy usage.

The Bearer Channel: How the Messages Get to Phones

The launch of WAP services followed by 2.5G services is taking place as the mobile phone industry is going through a period of consolidation. The term 2.5G is used to describe the technologies being introduced on the evolution path from *second generation* (2G) to 3G systems and includes *High-Speed Circuit Switched Data* (HSCSD), GPRS, and *Enhanced Date Rate for Global Evolution* (EDGE).

Circuit Switched Versus Packet Switched

Behind the networks, portals, applications, and services is a mountain of infrastructure supporting content delivery, data compression, device and network interoperability, and billing. The need for this infrastructure is due to the complexity of the U.S. landscape, with numerous wireless transport protocols (such as CDMA, GSM, and so on) all trying to move content sourced in a variety of different formats (such as HTML, Unix, Linux, NT, C++, POP3, IMAP4, and so on) to devices with various capabilities and operating systems (such as Palm, PocketPC, EPOC, *Research in Motion* [RIM], Stinger, and so on). The wireless gateway infrastructure can sit at the content source, the network operator, or be hosted by a wireless *application service provider* (ASP). In any event, the move away from the disparate content protocols (such as HTML, WAP, and i-mode) to the X-HTML standard should benefit all wireless data users.

The major challenge for carriers will be the engineering, billing, and back-office infrastructure move from circuit-switched voice networks to packet data networks. Traditional planning tools for telephony (voice channels and blocking rates) are generally not applicable to packet-based networks, and there is not yet an established model for hybrid networks. Therefore, as packet data becomes a significant part of the network load and as voice telephony moves to *voice over IP* (VoIP), carrier-pricing models will evolve. This is further exacerbated by the asymmetric nature of Internet data usage and the QoS pricing issues this could raise. Carriers' current pricing is very simplistic: Use your per-minute voice rate and transfer data over it if you want to (instead of talking). The carriers can do this because the channel is tied up anyway. In the future, the key measure for packet data will be throughput (in other words, your speed on the street); it is the best common denominator that takes into consideration all aspects of the network and technology. It also takes into consideration user expectations about service.

Circuit-Switched Data

Circuit-switched networks are not optimal for wireless data usage as they require a constant connection between the network and user. Circuit-switched services are acceptable for lengthy communications such as fax or video. Using circuit-switched services results in charges based upon per-minute usage and potential network bottlenecks. Most nonphone devices do not operate over these types of networks and many WAP-enabled phones are not able to keep track of time spent retrieving data.

Packet-Switched Data

Packet-switched communication is best used to support data transmission that is bursty, such as messaging applications. Packet-switched networks such as *Cellular Digital Packet Data* (CDPD) and private packet radio networks such as those from Cingular Interactive's Mobitext network (formerly BellSouth Wireless Data) and Motient's DataTAC network, and Aeria Network's Ricochet transmit data to the end-user device in packets. Users are typically charged for these networks on a data-load basis, which, although difficult to record on most mobile phones, assures users that they are only being charged for what they get and not for time spent entering data or navigating. Packet-switched services are optimal for accessing information over IP-based networks. Networks based on 2.5G technologies such as CDMA 1XRTT and GPRS are packet-based technologies. Carriers began deploying these technologies in 2001 for service delivery in 2002.

Almost all applications currently envisaged (even video applications on PDA-sized screens) do not require higher bandwidths than those that will be available with technologies such as EDGE. This technology is far less expensive than 3G since it requires only an upgrade of the existing GSM network rather than the deployment of a new one. Yet, mobile operators have invested billions of dollars to obtain licenses and deploy 3G networks (because mobile operators require more spectrum) for two major reasons. First, they want to set up operations in countries where they are not yet present.

This strategy requires a license unless network operators want to become virtual mobile operators. Second, many mobile operators will face saturation problems on their networks and therefore need additional spectrum for additional capacity in their own country. The debate over the bandwidth possibilities offered by 3G is actually a debate on the number of users rather than the bandwidth requirement per user.

Details by Market Region

At the beginning of this chapter, we stated that regional requirements and differences create unique market environments for wireless messaging products and services. These differences create environments where some products are more suitable than others. In the following sections, the regional basis will be examined.

Japan's NTT DoCoMo i-mode Service

The most talked about wireless messaging services comes from Japan. Since its launch on February 22, 1999, NTT DoCoMo i-mode service has over 32 million total users (including e-mail). The *i* in i-mode stands for *interactive*, *Internet*, and *independence*.

i-mode is now the largest *Internet service provider* (ISP) of any kind in Japan. Millions of Japanese are getting their first Internet experience on their i-mode-equipped wireless phone, which is similar to the first messaging experience via SMS in Europe and via the traditional PC in North America. i-mode's unique features include a trouble-free display of text and color graphics, constant connection, and thorough content including e-mail, financial services, games, and news. i-mode charges a flat fee of 300 yen (approximately $3) per month in addition to 0.3 yen per 128-bit packet. All the technological innovations are important, but the societal issues in Japan must also be considered. For example, the installation of landline Internet services in Japan costs approximately $500 per household. The average commute entails one to one and a half hours of travel, which is typically on a train, providing plenty of free time to communicate, coupled with the relatively inexpensive cost per packet. The Japanese language is very precise, which readily lends itself to device messaging, and involves less ambiguity in interpretation than the English language. This environment, united with attractive rate plans, is a catalyst for strong wireless messaging and content usage.

To achieve this feat, i-mode has changed the business model in which users were charged for information—by the amount of information downloaded as opposed to the user's duration online. This business model has created another avenue for DoCoMo to attain revenue aside from the traditional voice service provider. As the first wireless web service, DoCoMo also achieved the first m-commerce capabilities—the capability for users to purchase items online (which is identical to the wireline PC's e-commerce).

The Triumph of Messaging—Two Stories

Messaging has triumphed in Japan and the Philippines, which have two different economies and social cultures. Japan is an affluent country, whereas the Philippines is seeking to increase their macroeconomic climate. So, how can two different societies drive the largest amounts of messaging in the world?

Before we can answer this question, it is important to identify some unique usability facts. Japan is an affluent country with larger per-person disposable income amounts, particularly when compared to other Asian countries, notably China, India, and the Philippines. Messaging, specifically e-mail, has succeeded in Japan because of a few primary reasons. The Japanese spend approximately 2.5 hours daily travelling to and from work, voice conversation in a bus or train is considered impolite, low PC penetration often makes i-mode the first messaging and Internet experience for many Japanese, and the technology provides always-on connectivity (similar to corresponding via SMS) and inexpensive per-packet pricing (purchase what one uses). These factors compounded with higher-than-normal amounts of disposable income increase the adoption of this technology.

This is quite contrary to the catalysts in the Philippines. This country is quite poor and the Japanese dwarf the level of disposable income found in that country. SMS, as opposed to e-mail, is the primary communication medium for Filipinos. As in Japan, pricing is also a key driver in the Philippines. About half of all messages sent are free. Globe Telecom, a Filipino mobile operator, offers 500 free SMS per month. In addition to those 500 messages, the primarily prepaid subscriber base sends approximately 200 additional paid SMS per month. This pricing model makes SMS attractive compared to cellular voice calls. When comparing SMS and voice pricing, a voice minute is approximately four times the cost. However, SMS has not cannibalized voice service. SMS has become a cultural phenomenon in the Philippines and has become another mode of communication and entertainment. SMS adoption seems to come naturally in the Philippines because language is not a barrier to entry and the population is quite young. About 48 percent of Filipinos are between the ages of 15 and 45 years old.

The European Union's (EU's) Messaging Addiction

SMS, as opposed to wireline Internet e-mail access, was the first messaging correspondence for many Europeans. The success of SMS in Europe is also partly due to societal issues. European youth, the primary drivers of SMS, use the service as a less expensive method of correspondence. The explosion in SMS usage in Europe and e-mail in Japan caused the pent-up demand for simple and affordable two-way text messaging (an alternative to more expensive voice correspondence) rather than any clever marketing or technology enhancements on the part of the network providers. In the Americas, the ambiguity of the languages spoken and the high PC penetration make SMS messages less than ideal.

Approximately 90 to 95 percent of worldwide SMS traffic is social text messaging, primarily because of very high mobile phone penetration in the youth segment enabled by prepaid services. Teens drive SMS usage, although the enterprise market is also now beginning to embrace the technology. Additional SMS drivers include early two-way availability and limited paging penetration (approximately 3 percent today). The remaining 5 to 10 percent is allocated to information services. Other uses for SMS include

- Notifying a mobile phone owner of a voicemail message
- Notifying a salesperson of an inquiry and contact to call
- Notifying a doctor of a patient with an emergency problem
- Notifying a service person of the time and place of his or her next call

Furthermore, the EU's mobile phone penetration is on average approaching the range of 75 to 85 percent, unlike the United States where mobile phone penetration just recently surpassed the 46 percent mark.

South America, South Africa, Australia, and the Pacific Rim

SMS usage in South America is just beginning to take form. Brazil holds the most promise. The economic conditions of this region are not the best catalyst for wireless service, as the amount of disposable income toward wireless usage is limited. Although the region is limited, SMS offers a less expensive method to communicate. Similarly, SMS usage in Africa is also limited since wireless network towers are not widely deployed.

Over the last two years, Chinese wireless businesses have defied economic gravity. From a total of just 50 million in July 2000, mobile subscribers in China grew to 117 million by the end of June 2001. They now exceed the number of subscribers in the United States, making China the world's largest market, with annual subscriber growth rate exceeding 10 percent. China represents a wireless market with bountiful opportunities, especially in the messaging arena. Many carriers in China share the optimism. China Mobile predicted that there would be 10 billion SMS messages sent at the end of 2001 and expects SMS to bring operators $121 million in revenue. More than 500 small startups have also been launched to provide services for China Mobile customers since SMS was introduced in 2000. Hundreds more are expected as GPRS and MMS take hold over this year or next. The Japanese market, primarily youth, will increase SMS penetration slightly over the short term, as the demand for wireless text correspondence is great, before declining with the advent of wireless IM.

Eastern Asia, the Middle East, and Africa will continue to increase penetration based upon the demand for an alternative to costly voice correspondence and the service appearing as a novelty in many emerging markets. These markets are not deployed, but when they do become more mature, SMS will be a mainstay application.

The Americas (includes N. & S. America)

The Americas will not experience the same wireless text-messaging usage or carrier revenue percentage as Europe, primarily due to just-provisioned interoperability, low penetration of SMS-capable handsets, high PC penetration, and the fact that competing technologies such as wireless e-mail have already begun to penetrate the market. The following are some reasons why North America has not experienced as high a penetration as Europe:

- North America has high PC penetration, with faster network speeds, large color screens, and keyboards.
- The time away from a computer is short; people usually have a computer at work and at home, reducing the need for SMS while traveling (there is no value in SMS, unlike wireless e-mail, which is linked to desired content).
- Inputting text is difficult.
- Free local telephone calls by network providers have spurred home landline Internet usage.
- The reliability of wireless networks and service is poor resulting in extended periods of no service.
- SMS roaming is limited.
- North Americans prefer voice interaction on a mobile device (voice communication is quicker).

- Wireless e-mail services from Blackberry and ViAir are available and are continuing to gain penetration.

MO capabilities were recently introduced in the Americas, complementing the already-available MT, although both are still in their infancy. North American network providers have been criticized for their blasé SMS offering, which is partially due to their insufficient capacity.

Carriers are hesitant to market SMS solutions due to their lack of wireless data competencies, insufficient capacity, regular service outages, and slow delivery speed. Marketing unreliable and limited SMS services while constantly comparing their sub-par offering to their European counterparts would create a humiliating situation for North American network providers. Hence, the network providers have limited offerings and will continue to be very unhurried to enhance these services.

Hype can be blamed for the user's expectation of a wireless device. U.S. carriers have been marketing wireless Internet as exactly that —the wireless Internet. Hence, users in the United States expect their experience to be almost identical to the wireline Internet and are disappointed with small screens, difficult text input, tariffs, and slow network data rates. Subscribers consider their wireless device (commonly a handset) to be companions, not tools, like a PC.

As subscribers and technology advance, another mechanism for unsolicited message spam increases. Wireline Internet users for years have been complaining about spam. Consumers now are concerned about spam sent to their wireless devices and the U.S. government has taken notice.

Messaging Legislation—Spam As wireless messaging systems become more prevalent, it is only a matter of time before spam—unsolicited electronic messages—becomes a nuisance for users of wireless devices. Many wireless providers get phone numbers in blocks of a thousand, and because the messaging address is the number of the phone, if you know the prefix of the provider in an area, you can easily access all of the numbers between 0000 and 9999 that have the same prefix. The U.S. Congress has therefore introduced legislation to protect end users. The Wireless Telephone Spam Protection Act, introduced by Rep. Rush Holt (D-New Jersey), would make it illegal to use any mobile telephone messaging system to transmit an unsolicited message. The bill, an amendment to the Communications Act of 1934, also states that the FCC may not exempt carriers that make calls that violate the prohibition. The proposed legislation aims to address wireless spam before it becomes a significant problem. According to the *Coalition Against Unsolicited Commercial E-mail* (CAUCE), the proposed legislation seeks to protect wireless messaging systems from suffering the same fate as e-mail, which is often abused by advertisers using spam.

This amendment would "prohibit the use of the text, graphic, or image messaging systems of wireless telephone companies that want to transmit unsolicited commercial messages." Ultimately, the bills would make it a criminal act to send a solicitation to a wireless device without the subscriber's permission. Subscribers would be allowed to sue companies that ignore their request to be taken off the mailing list. If end users are not protected by federal regulations, commercial message senders and network providers will be forced into a legal debate over freedom of speech versus privacy of customers.

Holt's wireless phone bill, referred to the House Committee on Energy and Commerce, specifically prohibits the use of wireless phones' text, graphic, or image messaging systems to transmit unsolicited commercial messages. It further states that "consumer protections must keep pace with advances in communications technology to ensure protection of privacy and personal time; and to protect the privacy of wireless telephone subscribers."

In the United Kingdom and EU, there have been some regulatory steps taken that have some government- and industry-backed initiatives on the books to secure the privacy of mobile phone users—both for voice and SMS messages. It seems that the United Kingdom gives a great deal of scrutiny to these types of issues and there is likely to be heavy regulation of SMS-based marketing. For instance, the United Kingdom's Data Protection Commission has sought to make it illegal to use wireless location technology for advertising purposes, even if customers choose to participate.

Japanese wireless carrier, NTT DoCoMo, the provider of the i-mode mobile phone Internet service, has launched an attack against spam. The company's most recent efforts to address the problem include blocking bulk spam from entering the DoCoMo system, providing DoCoMo's i-mode customers with new ways to block spam, taking additional legal action against companies sending junk e-mail, and reinforcing public information efforts.

DoCoMo will invest 1 billion yen ($8.22 million) in systems to block unwanted e-mail. Some 950 million e-mail messages are handled daily by the company, and about 800 million of those are returned to senders because of invalid addresses. To eliminate the heavy burden that bulk spam places on the company's overworked i-mode network server, DoCoMo requested permission from Japan's Ministry of Public Management, Home Affairs, Posts, and Telecommunications to block any e-mail sent to large numbers of invalid e-mail addresses.

The measure also would inhibit spammers' ability to create lists of valid e-mail addresses, pointing out that some spammers send bulk e-mail using computer-generated lists. Valid addresses are accumulated through a process of elimination, since any invalid address that triggers a user-unknown message from the i-mode server is dropped from the list. With the new blocking function, however, mass spammers will no longer be able to receive these messages. Until now, DoCoMo's anti-spam efforts have included taking legal action against spammers and encouraging users to change their e-mail addresses via their handsets in exchange for 400 free packets (worth 120 yen) of data communications each month. Customers are asked to implement usernames that contain random alphabetic and numeric characters, instead of using their phone number as their default username. This practice makes it easier for spammers to bombard large numbers of people. Regardless of the measures carriers take, the end result needs to be the same—inhibit spamming.

Mobile Device Overview

Now that the formats and regions have been discussed, the following question arises—what type of device do you need/should you use?

The device form factor is arguably the most significant catalyst for the development and progression of the wireless messaging market.

Device form factors are critical because they define the user experience. If the device form factor is unpleasant or difficult to operate, not only will users not operate the messaging and all other applications, but the device will also sell. People will remain free to select their device of choice; many users will carry at least two (voice and data specific—for example, a Blackberry for messaging and a small discrete phone for voice). A converged device may not become a category killer. On the contrary, different devices will appeal to different people (Ford versus BMW versus Jeep) and be tailored to a specific need.

When considering a new wireless device, the following items should be considered:

- What wireless network does it work on?
- Will service be most useful on a phone, PDA, laptop PC, pager, or youth-oriented device?
- If more than one device is needed, how complicated is synchronization?
- Is there enough memory and processing power to support the desired applications?
- Can applications be installed wirelessly or via PC connection, or is the device closed?
- Is there a large software development community for the device's operating system?
- Is the device supported by an employer's IT department?
- Based on evolving network standards, what is the life expectancy of the device?

The answer to these questions will determine whether or not someone needs a Kyocera Smartphone, a Palm Pilot, a Blackberry with voice capabilities, or a Compaq iPAQ to run Excel, Word, and Outlook. Some users may prefer the simplicity of a Palm, whereas others prefer a phone but want data access. A heavy message user may need a QWERTY keyboard such as the Blackberry. Someone else may want speech-to-text capabilities, and a teenager might be content to tap on a numeric keypad. In any of these situations, there is no correct device. The only sure thing is that data-centric devices are going to proliferate on the carriers' networks. To capture the traffic and service of these users, carriers are transforming themselves to offer services, such as faster data transmission speeds, capacity, and customer service for a completely new device group that requires a GUI and content other than two people speaking to each other.

Smartphones

Industry observers believe the wireless domain is at the onset of an explosion in enhanced wireless devices, specifically smartphones. A smartphone combines voice and PDA capabilities. Despite the rapid rate at which device convergence is occurring, a smartphone can still be defined as simply any voice communications device that also can handle data. In short, it could consist of a portable voice radio and miniature computer combined with an operating system that runs programs that can be dynamically updated and managed via the radio link.

Being touted as the ultimate productivity tool, smartphones are comprised of everything from address books to *Personal Information Management* (PIM) software to the Internet. Activity is not limited to traditional wireless behemoths such as Nokia, Ericsson, and Motorola, but it extends to Palm Computing, Alcatel, Sony, Mitsubishi, and IBM.

The smartphone market has seen large, clunky combinations of phones and information appliances give way to sleeker, thinner phones designed to capture mass-market appeal. New smartphones offering simplified e-mail, calendar, and contact applications will gain significant popular acceptance in the future primarily because of their all-in-one capabilities. Estimates of smartphone growth vary widely from IDC's estimate of 12.9 million shipments in 2003 to Psion's estimate of 78 million in the same timeframe. According to Psion, application-enabled smartphones and communicators are likely to outstrip basic and browser-enabled handsets by 2005.

One of the keys to determining the viability and penetration of WAP and other data-oriented services will be the pricing of these new handsets. Thus far, available smartphones are priced at the high

Figure 3-6 The Sendo Stringer, Samsung, The Kyocera, and Handspring Smartphones (from left to right), which use the WindowCE, PocketPC, and Palm OS, respectively (Source: Kyocera, Samsung, Handspring, and Sendo)

end of the market—some above $700. Throughout 2002 and 2003, manufacturers plan to web-enable existing favorites at little to no incremental cost, as the browsers are essentially free, and subsequently decrease smartphone prices. Greater handset availability, model choices, and more accessible prices (near $100 to $400) should act as a terrific stimulus for industry growth. Figure 3-6 shows pictures of some early smartphones commercially available in the market place.

PDAs

Since the introduction of the Apple Newton in 1993, PDAs have gained momentum as organizers and small complements to the PC. More recently, the PDA space is repositioning to include advanced communications, including Internet access. Several newer models feature access to electronic messaging, faxes, and interoperability with office *local area networks* (LANs). The PDA is considered a natural market for many early wireless data applications due to its strong customer base of mobile professionals and its enhanced processing power, memory, and screen size, which are all greater than current and planned data-enabled wireless phones.

An increasing range of data-enabled handheld devices will spur additional and more compelling handheld applications, service offerings, and adoption. According to IDC projections, the PDA and other handheld markets will rise to nearly 19 million by 2003. Although this presents a significant market opportunity for wireless data solution vendors, it is a far cry from the 1 billion or more wireless phone users projected during the same time period. Several companies, including IBM, Sony, Hewlett-Packard, and Compaq, are developing new handheld offerings.

Recently, a number of PDA manufacturers have begun incorporating cellular radios in their devices. This enables the PDA to perform all the functions of mobile phones and then some. Figure 3-7 shows examples of traditional PDAs.

Messaging or Paging Devices

Messaging devices can seamlessly send and receive text messages primarily because of their integrated wireless modem. Whereas two-way messaging networks are limited by bandwidth constraints for the amount of information that can be sent over their networks (typical data rates are 14.4 Kbps, which is similar to early dial-up modems for PCs), the principal intent of these devices is for messaging, not browsing. However, they do offer benefits over cellular, *personal communications services* (PCS), and other networks since their coverage is typically greater and they offer an always-on capability

Figure 3-7 The Handspring Visor running the Palm OS and the HP Jornada 720 running Microsoft Windows for Handheld PC 2000 (Source: Handspring and HP)

for receiving and storing messages. Two-way messaging devices can be used over ReFlex networks, narrowband PCS, private packet radio (such as Motient or Cingular Interactive network), and *wireless area network* (WAN) packet data networks. Recently, several Internet providers such as AOL have introduced two-way devices, which can be used to access AOL e-mail and IM services wirelessly.

There are other benefits offered by two-way messaging devices over mobile computers. Typically, they have greater battery life—as much as two weeks between battery changes for some devices under moderate use conditions. Also, because they are constantly on, they can receive messages instantly rather than having to wait until the user tells the device to query the network for new messages.

A definite trend toward device convergence for most of these pager-like devices is shown in the devices listed in Figure 3-8. Besides their normal data capabilities, these devices have a built-in voice radio and come with an earpiece and microphone combination. Motorola was first to market with its Accompli product. Figure 3-8 provides pictures of data-centric devices incorporating voice functionality.

Although the paging industry has been in decline, 1.5 text (message notification) and two-way messaging (message response) are

Figure 3-8
The RIM 8510 and the Motorola Accompli (Source: RIM and Motorola)

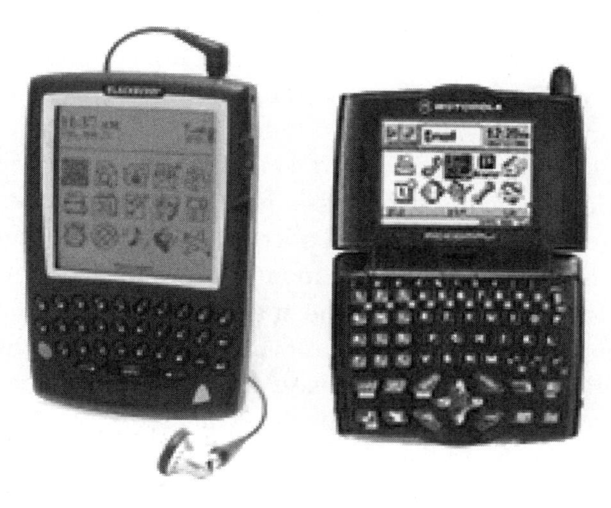

seeing growth. Forecasts from IDC predict 97 percent annual growth in two-way messaging through 2003. The advanced messaging market will enjoy as widespread adoption and subscriber rates of the smartphone market over the next several years. Thus far, the RIM Blackberry is available, which is arguably the most innovative messaging product to come along in years.

Messaging Using These Devices

Messaging via traditional PDA devices typically requires purchasing a modem for wireless access. Palm later introduced the Palm VII, which incorporated an internal modem. These modems typically support IMAP and POP3 e-mail. Recently, developers have made Exchange and Notes e-mail support commercially available for enterprise users, which should lead to more widespread adoption. Wireless ISPs for PDA devices typically offer two pricing plans: a limited usage amount for a set fee and a nominal fee for usage above the set amount, and an unlimited amount of usage for a premium amount.

Messaging via the smartphones today is priced as if the user has initiated a voice call (a WAP session), although this will change as packet networks become more widely deployed. These devices offer a similar WAP experience to the small WAP-enabled phones, but with a larger screen combined with PIM functionality. The user of the address book can typically begin composing an e-mail directly from two or three screen taps. Additionally, smartphones with improved technology can locate and hold network coverage better when their smaller, less sophisticated counterparts lose coverage.

Messaging via two-messengers has evolved over the last two years from mere beepers into devices that in some cases border on usurping the functionality of mobile computing devices. Several of the messaging devices on the market have full QWERTY keyboards, offer PIM functionality, and have manufacturers who are creating applications for the devices. In some cases, customers find themselves questioning whether they need to purchase a mobile computing device in addition to their messaging units.

Getting Around the Text Input Problem

For the devices described previously, or for any small wireless device that requires the entry of textual information, typing in a message can, in the beginning, present complexity and frustration. A number of innovative solutions have been brought to market to try to make this issue simpler.

For example, Tegic's T9 predicted typing technology is one solution. After a single tap to a key in sequence is made, T9 does its best to figure the word that the person was typing. This approach was not fail-proof, but it marked a vast improvement to what came before. Early on, for cell phones, for example, a text message could only be entered by triple tapping (hitting a single key repeatedly until the number, letter, or symbol you are looking for is displayed). This was the only way that people could enter textual information.

Today, new and innovative methods for text entry to mobile devices can be seen on the market. One example of such a new technology can be found by looking at new products. The industry has recreated device keypads by integrating an alpha keypad into an old 12-digit numeric keypad. Figure 3-9 shows an up-close photo of the popular QWERTY keyboard.

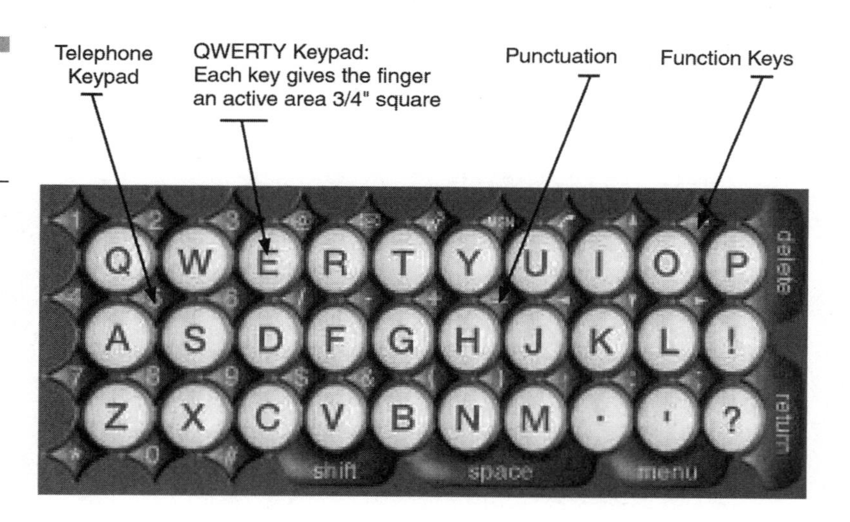

Figure 3-9 The QWERTY keypad (Source: RIM)

Another area of current innovation for text entry to wireless devices is handwriting recognition coupled with Bluetooth wireless connectivity to the user's device. This looks like a small pen or stylus that gets connected via a Bluetooth to the wireless device. Through this Bluetooth connection to the device, the message is typed in as you write on a pad of paper with your stylus. Figure 3-10 illustrates a Bluetooth pen.

The main driver of wireless data and the capabilities for success are here today. 2.5 and 3G networks are not needed for the success of messaging, which has already been established in Japan and Europe. The device replacement market presents a greater barrier than network speeds. Will users replace their current devices to use enhanced messaging? Only time will tell. As more users demand new features, devices will look for growth among mainstream users. The key will be delivering devices that have just the right mix of voice and data functionality at an affordable price for both the device and ongoing services.

Figure 3-10 Anoto Stylus input devices

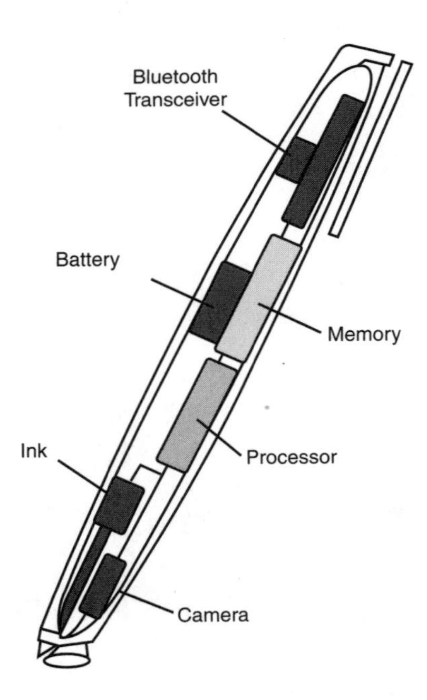

The following main theme can be taken away from this chapter: The messaging format and device people use is unique to their requirements and social patterns. But what are those application requirements? What are all the applications that one can use via messaging? The next chapter will discuss those applications that may influence the service and device purchased.

CHAPTER 4

Applications of Wireless Messaging

This chapter discusses the wireless messaging applications, and the goods and services that have driven and will continue to drive the escalating growth of mobile messaging. To start with, it is important to establish what makes a wireless messaging application and where the value is for companies competing within this space. A survey of the applications available via wireless messaging is provided in this chapter to address those goods and services designed for personal and professional mobile users as well as the potential role for wireless devices in mobile payments and wireless commerce.

From Person-to-Person (P2P) Messaging to Content and Services

The progression of technology behind the success of the wired Internet has repeated itself in the wireless world. Just as the Internet started with e-mail, the wireless Internet has started with messaging. In both cases, *person-to-person* (P2P) communication has turned out to be the keenly sought killer application for boosting data traffic and maximizing revenues.

The other driver for Internet data traffic has been entertainment services. As the P2P messaging market matures, the players in the wireless industry are starting to realize the enormous revenue-generating potential of mobile content. Recent reports by the analyst Jupiter MMXI reveal that by 2006 spending levels for mobile content will reach figures of around $3.3 billion in Europe alone. This compares to the equivalent figure of $1.7 billion on PCs. With these estimated revenues, it is not surprising that handset vendors, mobile network operators, and content providers are banking on wireless content to stimulate demand in the mobile sector.

Following i-mode's Lead

The runaway success of NTT DoCoMo's wireless Internet or i-mode service has provided an enviable business model for content provision over wireless networks. Since its launch in February 1999, the i-mode service boasts well over 34 million active users, a figure that increases at the rate of around 50,000 new users each day.

Much is made of the difference between i-mode and the *Wireless Application Protocol* (WAP), but it is important to remember that unlike WAP, i-mode is a user-focused service that combines a solid business plan with existing technologies. NTT DoCoMo recognized that the key to its success was to attract users to the service. Technological advance and innovation only attract the first movers. The compelling services and attractive content bring the hoards. The key to i-mode's success was threefold:

- *Develop minimum specifications based upon service concepts.* Once these were in place, DoCoMo worked with manufacturers to provide the devices that users would want to use to access the Internet, featuring large screens, color graphics, and polyphonic ring tones to pave the way for rich content.
- *Act as the conductor as well as the conduit.* NTT DoCoMo made the decision to act as the *Internet service provider* (ISP) for i-mode users by providing access to both the Internet and an e-mail account, enabling subscribers to communicate with other mobile phones and PCs anywhere in the world.
- *Act as the gateway to the mobile Internet.* DoCoMo set itself up as the portal to the i-mode service and, in support of this, acted as an aggregator of rich content from a wide range of content providers. Recognizing that this walled-garden approach would prove too restrictive eventually for the more sophisticated i-mode users, NTT DoCoMo enabled i-mode to access any i-mode-formatted Internet site. There are currently over 30,000 i-mode-compatible sites.

What Happened to WAP?

Unlike solutions such as Internet e-mail and *Short Messaging Service* (SMS), WAP is a person-to-machine service, not a P2P service. Slow connection times, a delay in the supply of WAP handsets, misdirected marketing, and the lack of attractive content all contributed to the perceived flop of WAP. After having poured so much energy into WAP, the success of SMS, particularly given the lack of marketing, took operators by surprise and is the reason many still do not have coherent SMS strategies. Operators are now turning their attention to combining the two solutions in a complementary fashion, where SMS acts as the bearer technology for WAP.

With the wireless medium attracting content providers as a new channel to sell and market their product's games, information services, and advertising, the market is set for another phase of exponential growth in SMS. The arrival of enriched messaging technologies such as *Enhanced Messaging Service* (EMS) and *Multimedia Messaging Service* (MMS), as discussed in Chapter 5, "Wireless Messaging Today," will enable users to further personalize their messages with pictures, sounds, and animations.

Messaging in the United States

With the predominance of *Global Standard for Mobile* (GSM) technologies in Europe and Asia Pacific, the enormous growth in SMS and the imminent development of EMS and MMS is great news for those two regions, but what about the United States? The United States and Canada are the only countries in the world where PCs outnumber mobile phones. This imbalance is caused by a number of factors, not the least of which is the cost differential between fixed-line and mobile calls. In Japan, expensive Internet access rates drove users to turn to the wireless world; in North America, however, the mobiles have priced themselves out of the market as local calls are free and PCs are cheap.

One fact remains: The mobile Internet, in one form or another, generates enthusiasm throughout the developed world, much like the Internet did in 1995. It is clear that both personal and professional consumers will push for mass adoption of this new technology over the next few years. Wireless messaging is the driver for this larger goal today, and although it is reaching levels of maturity in Europe, messaging has a long way to go in the United States.

What Makes a Wireless Messaging Application?

The mobile Internet differs in three main ways to the fixed Internet:

- *A mobile device is very personal.* Unlike the PC, the mobile phone is likely to be owned and used by one person and be carried by that person or kept within reach all the time. Whereas e-mail is delivered to a machine sitting on a desk, text messages go directly to the mobile phone's user.
- *Operators can tailor devices.* Mobile network operators can determine the menus and services that are specific to a user's mobile device. This element is not possible with desktop PCs as the users have far more scope to play around with settings.
- *Consumers expect to pay for mobilization.* Users recognize that mobiles cost money and that the network operators have a channel by which they can charge a premium for the privilege of accessing services while on the move. PC e-mails or instant messages are essentially free. Even when text messaging is at a premium, such as while roaming on international networks overseas, mobile users are still prepared to pay for the convenience.

With respect to these unique properties, the value proposition of wireless applications is dependent on both location sensitivity and personalization, which is enabled through profiling and portals. By complying with these two key attributes, applications will be useful, convenient, and personalized, and therefore will go on to leverage the core attributes of mobility by being time sensitive, location sensitive, customized and filtered, optimized for mobile device interfaces, secure, and easy to use. All of this combined with the consumer's willingness to pay ensures that true wireless applications can expect a fantastic result and *return on investment* (ROI).

In addition to dealing with the positive characteristics of the wireless device, it is also essential to recognize the drawbacks. Mobile devices have a slower connection speed, limited screen size, and keyboards of minute proportions as compared with the PC. It is vital to learn from the mistakes made with WAP where technology and marketing hype led people to expect the wireless equivalent of browsing from a desktop by recognizing these limitations and tailoring applications to offset these negative points.

Where Is the Value in Wireless Messaging Applications?

What Is the Value Chain?

The value in wireless messaging applications comes from the association of different ingredients in the value chain that are essential to the implementation and deployment of a wireless service. Figure 4-1 illustrates the wireless value chain.

One end of the chain consists of the disparate world of wireless operators, which has on average three to four wireless operators per country. The other end consists of the content/service providers targeting audiences at local, regional, and global levels. Each component of the chain combines to solve a mix of technical, marketing, financial, and creative issues to enable the successful deployment of wireless *value-added services* (VASs). The different players in the chain can find themselves in different business relationships when tackling this challenge and can each simultaneously act as partner, competitor, and customer/supplier to a single company. This is common to the wireless environment since it is large in terms of vertical

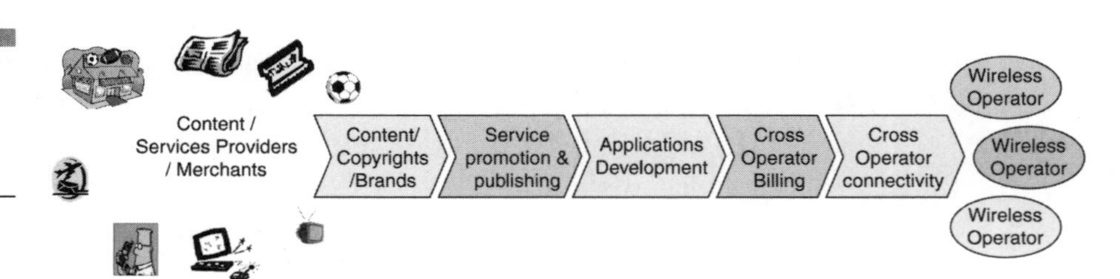

Figure 4-1 The wireless value chain (Source: Mobileway)

markets, the range of services, and the geographical presence of the companies within it.

Revenue Models

Revenue is generated via two main approaches: the publishing approach and the bill-on-behalf approach.

Figure 4-2 depicts the publishing approach. With this approach, content/service providers bear the costs associated with the deployment of the wireless services and pay the wireless operator to use its network infrastructure, a cost that is considered a publishing fee. In this case, content providers generate revenue directly by adding the cost of publication to the price of the good or service offered to its customers or by indirectly using the wireless service as a means to promote the product through direct marketing, advertising, and so on.

Figure 4-3 depicts the bill-on-behalf approach. Using this method, the end user orders the service directly from the content/service provider and is charged for it on the phone bill. A share of the amount charged by the wireless operators is passed onto the content/service provider—this approach is known as *revenue sharing*. Various business models featuring diverse revenue allocation agreements can be built to reflect the various roles and value

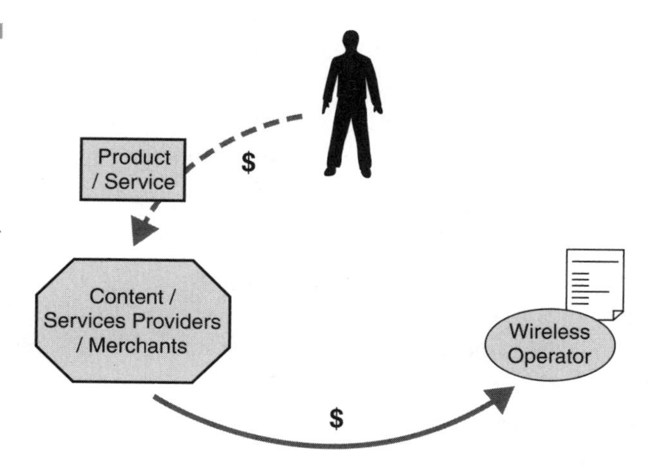

Figure 4-2 The publishing approach to wireless messaging revenues (Source: Mobileway)

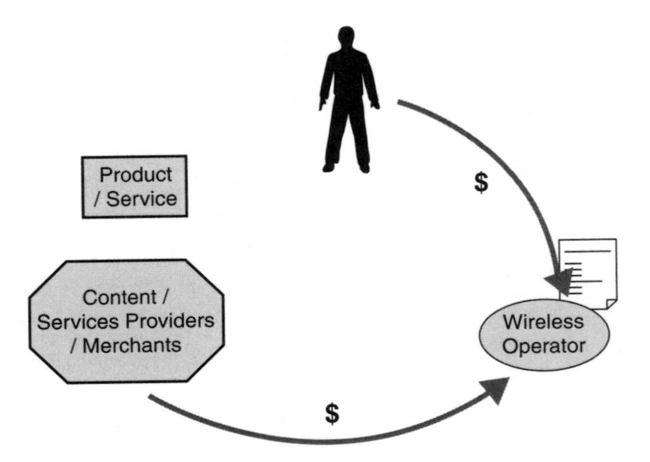

Figure 4-3 The bill-on-behalf approach to wireless messaging revenues (Source: Mobileway)

brought by the different players in the value chain. There are three main models:

- **Model 1** Content/service providers have to negotiate revenue-sharing deals with each individual wireless operator and subcontract service implementation or good deployment to the appropriate third parties. This model makes deployment very onerous, particularly when money must be collected and allocated to the various parties.
- **Model 2** A third party aggregates the operators and all other activity in the value chain to provide a single interface for content/service providers. This hides the complexity and heterogeneity of the process from the content/service providers by providing them with a seamless deployment. In this case, revenue flows from the wireless operator to the third party and then to the content/service provider.
- **Model 3** In this case, the third party is no longer a single entity, but is now a group of different parties cooperating to combine the various expertise necessary in the value chain and provide the complete solution. The different parties could be a group related to the technical implementation led by the connectivity and cross-billing provider, and a group related to the promotion and publication of the service led by the content provider. In this scenario, revenues flow from wireless operators

to the entity providing the connectivity and billing, and then to content/service providers.

What Are the Challenges Faced by the Various Parties Involved in Implementing Wireless Applications?

Technical The connection to the wireless operator environment needs to handle two major functions: delivery and billing. Specific connections need to be established to cope with the limitations of each module, such as the limited bandwidth within SMS-Cs or the inability to charge prepaid users who have exhausted their credit. Many of these issues still have no solution, forcing operators, content providers, and third parties to fix the problem in an indirect way, such as by regulation. Understanding these technical constraints and their origin is crucial for managing these limitations at the most appropriate place in the value chain.

Marketing The accessibility to chargeable services from the mobile device if they have not been adequately promoted can result in users unknowingly running up a large bill. The relationship between operator and content provider is crucial at this point to ensure that the content is suitable for the mobile user and that they maintain their respectability with their user base. This calls for a level of trust among the different parties.

Legal The main legal challenge is due to the immaturity of the wireless services market and its lack of uniformity in the regulation area. In some regions, a service and approach might be appropriate, whereas in others, it is not. This is the case with issues such as *mobile spam* where regulation varies from country to country. Overcoming this requires a level of trust, good communication, and clear responsibilities within the value chain.

Financial Charging wireless services to the mobile user's phone bill introduces an element of financial risk and liability for operators and content providers. The time lag between service provision and the issuance of the bill to users is already around one month. Following the various accounting and validation operations by operators, any discrepancies in traffic levels and services delivered only now become apparent. In addition to this processing delay, wireless operators usually implement lengthy payment terms (sometimes approaching 120 days) before they pass on the share of revenue to content/service providers. This situation can cause a significant cash flow problem for content providers.

Challenges Met by Access Aggregators

If these issues are overcome, the development of wireless services becomes straightforward. In the short and medium term, this can only be achieved thanks to a trusted third party who can make sure that the interest of each party is well understood and fulfilled, and that the heterogeneity of the telecom environment is reduced and hidden to the content/service provider. This third-party role can be filled efficiently by an access aggregator. By easing these challenges, an access aggregator ensures that each party can concentrate on developing its own business while generating new revenue opportunities at the same time. The main access aggregation role is fulfilled when the company provides the following:

- Cross-operator connectivity and billing capability
- A portfolio of off-the-shelf applications (including support and customization)
- The management of provisioning and running of wireless service
- Cross-wireless-operator clearing and settlement facilities

This last criterion is the hardest to meet. Currently, Mobileway is the only company able to fulfill this role. Mobileway has leveraged its privileged relationship with Citigroup to address this issue and speed up the time that a content provider or merchant gets paid for their goods or services. Through this relationship and the arrangements in place with Citibank, Mobileway can not only secure revenue returns in less than 60 days, but it can also collect monies in 90 currencies and allocate funds in its choice of currency. This pioneering element in its business model, called *secured revenue*, is unique and expected to be essential for attracting both large and small content providers to move into the mobile space.

A Survey of Wireless Messaging Applications

Now that we have addressed the theory behind wireless messaging, it is time to look at the applications. Despite the blurring of the line in between one's workday and personal life, in their broadest terms, wireless applications can still be split into two main areas: personal and professional services. For the purposes of this study, payment applications have been tackled separately; this enables us to address micro- and macropayment in general without attaching them to any specific applications.

Personal Services

Information Services—Alerts and Notifications

In terms of market share, information represented the second largest category in 2001 with around 21 percent of the market; the largest category was music (according to ARC Group). The total number of users of wireless information services is set to grow 18-fold by 2006, making it the largest category. This large-scale adoption is due to the business rationale behind the provisioning of general information services. It is designed to add value to a service provider's package and therefore strengthen competitive offers rather than providing a direct content revenue stream. Mobile information services have become ubiquitous through their inclusion in many standard wireless subscription packages or as part of free content offered on web portals.

Current wireless information services comprise only pull services that are either paid for on a per-use basis or as part of a subscription. To be successful, wireless information services must provide content that maximizes the benefits of the wireless medium and is compelling enough for users to be willing to pay. As today's text-based messaging moves toward next-generation technologies, it is vital that information services evolve in line with these technological developments to offer a more interactive end-user experience.

Financial, Banking, and Stock Alerts Internet banking, online services, and brokering have already achieved great success and wielded an influence over a previously traditional financial fraternity. Major banks and investment institutions have either extended their banking activity to the Web or have streamlined their operations to leverage the cost-saving aspects of the Internet, enabling greater savings to be offered back to the end users. This has even led some banks to create a service solely on the Internet, granting today's customers complete control and flexibility to manage their assets. Recognizing the even greater potential offered by the wireless medium, banks and financial institutions continue to investigate the best way possible to adapt their service offerings to meet the constraints of the wireless channel.

This type of service is the most time sensitive of all mobile information services, making it potentially the most suitable for the mobile interface. The need to be accessible to receive notifications while on the move is crucial. The following are some services that are available to mobile users:

- **Stock quotes** Real-time quotes, current trading price with change on the day, high and low levels, and volumes traded.
- **Mutual fund quotes** Access on request to a range of funds local and specific to a user's market.

- **Market information update** Vision of the top stock gainers/losers on the key markets and real-time composite snapshots of the major indices volume with change on the day.
- **Local market update** Commentary and indices regarding a local trading market.
- **Portfolio** Any individual stock, mutual fund, or index can be tracked according to a user's portfolio and any movements can be highlighted.

Case Study: Citialerts—Wireless Banking in the Middle East and Asia Pacific

Who: Citibank, Mobileway.

Where: Singapore, Malaysia, Australia, Indonesia, Philippines, Hong Kong, Thailand, Taiwan, Guam, United Arab Emirates, and Egypt.

What: Citibank, which is one of the world's largest banking groups within Asia, has operated for the past 100 years within 14 countries and has an employee base of around 30,000. Citibank launched its alerts service in 11 countries to provide customers with the flexibility to track their bank accounts and receive the latest news and stock prices anywhere in the world automatically via their mobile phones or Internet accounts. Customers sign up to receive alerts related to their spending habits and withdrawals as well as the latest movements on the NASDAQ, New York Stock Exchange, and Singapore Stock Exchange. Alerts on up to 10 stocks can be monitored and triggered at any time by selecting a variety of options to monitor movements, such as rises above, drops below, percentage rise, percentage drop, yesterday's close, and so on. Customers can also choose the frequency and type of alerts.

Citibank uses the mobile transaction leader Mobileway to ensure seamless delivery in real time of the SMS-based alerts via any host operator across the region. Time is critical to ensure the success of the service.

Why: To arm Citibank's vast customer base with information needed to make quick and sound investment decisions.

Outcome: Launched initially on Singapore networks only, this free alert service has proved so popular it has now been rolled out to 11 countries worldwide and could possibly be extended further. To ensure initial success, users were offered an incentive to use the service. New subscribers were given the chance during the three-month launch period to win a holiday for two to the Maldives.

Sports, News, and Weather Updates Sporting updates were among the leading drivers of consumer adoption in the early phases of the mobile services market development cycle. They tap into the benefits of mobile communities by tailoring content according to existing fan bases. For example, Manchester United Soccer sup-

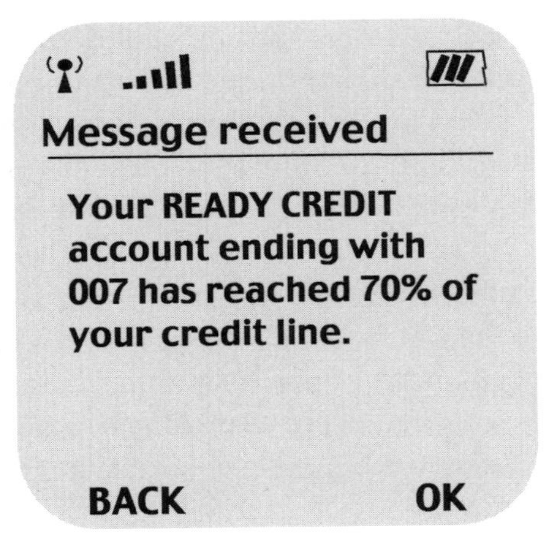

Figure 4-4 Citibank alert sent to a user's phone (Source: Mobileway)

porters can sign up to receive the latest results, including real-time scores in any game where their favorite team is competing. This can be extended to include any news updates regarding players or transfers. Some even offer chatroom services to promote interactive text and voice messaging among the community members. The 2002 Football World Cup in Japan and Korea achieved record highs that were close to the level of SMS alerts sent within Europe. Operators, merchants, and other broadcast media encourage soccer fans to keep up-to-date with the latest on their national side or follow the fortunes of other teams in their pool.

News alerts have proved successful on the Internet. Many publications are adapting their service to send out daily news roundups inviting readers to access the site via a *Uniform Resource Locator* (URL) link to find out more regarding a story of interest. Some of these services have adapted their offering to the wireless user by inviting consumers to sign up to a news service on the Internet and set up individual parameters to tailor the service to suit particular needs. For example, users following the parliamentary elections in the United Kingdom could sign up to receive alerts when developments arise.

Mankind's obsession with the weather has long been noted, and many of the Web's most visited sites are linked to meteorological organizations for consumers hungry for predictions about the weather for the near future. Offering users the ability to receive daily updates on the weather has proven popular. Time is again a critical factor in this service; visitors to outdoor festivals or events, such as the notoriously soggy Wimbledon tennis tournament, need to know the possibilities of rain ahead of time to establish the suitable attire. Weather updates form a typical part of operators' or service providers' subscription packages.

Traffic and Travel Updates This is the area most likely to benefit from the developments being made in *location-based services* (LBS) because of the improvements in telematics and onboard vehicle computers geared toward this service. However, the simple text-based *infotainment* side of traffic and travel has proved to be a key service to include in many operator subscription packages.

For example, these services give drivers the option to send a text message to request information from breakdown services to traffic congestion on a particular route or provide overseas travelers the chance to request flight arrival and departure information from a relevant airline.

The improvements in location-based technology will enhance users' ability to access specific information regarding directory listings of amenities, traffic black spots, or roadwork zones in relation to their immediate positioning. For further information on these developments, see the section "Location-Based Services (LBS)."

Music Information As developments in mobile payments move toward enabling users to purchase low-value goods including music CDs directly from the mobile handset, the scope for music information on demand is growing. Recent examples of interactive services such as those from Shazam Entertainment in the United Kingdom and MobiQuid in France enable users to identify a particular piece of music by dialing a service and receiving the background details to the track via SMS.

Case Study: Song Tagging with Shazam Entertainment

Who: Shazam Entertainment, Mobileway.

Where: United Kingdom.

What: Shazam has developed a real-time song identification service. When users hear a song they like, they simply dial Shazam's four-digit number from their mobile, let their phones listen along to the music for 15 seconds, and immediately learn the name of the song and the artist via SMS. The company has developed a breakthrough technology in audio pattern recognition to enable this service, for which six patents have been filed. Shazam's flagship product enables mobile operators to offer a music recognition service to all of their customers without requiring an upgrade in handsets or networks. After identifying the desired track, the tune is tagged (see Figure 4-5). Users can then

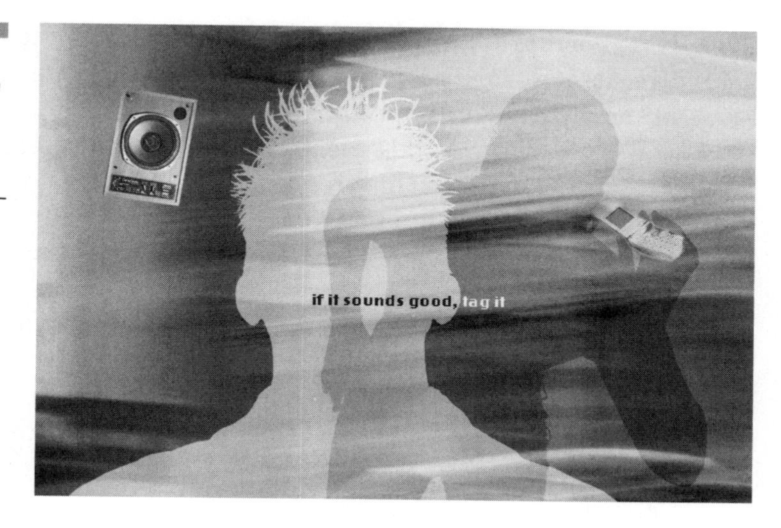

Figure 4-5 Tag it from Shazam Entertainment. (Source: Mobileway)

select from an array of options to interact with the song, including access to a personalized stored tag list on the Web, the ability to send a clip of the song to a friend, or the chance to purchase the single or album directly from the phone.

Why: To offer users the chance to identify any tune or piece of music playing in the background in a bar/cafe, on the radio, and so on.

Outcome: Service is scheduled to launch later in 2002.

Horoscope and Joke of the Day Jokes and horoscopes were among the most popular features of the early text-based SMS services. Horoscopes remain the most popular mobile VAS after e-mail in Italy and Singapore; mobile network operator M1 has more than 2,000 subscribers to its horoscope service. These services are extended in some countries such as Japan to include fortune-telling services, love horoscopes, and places where users can chat to potential partners fitting their astrological profile. NTT DoCoMo has been able to charge up to 100 to 300 yen for its fortune-telling content.

Text-based joke services have shown year-on-year compound growth of around 67.95 percent according to forecasts by the Arc Group. Although this is expected to experience a downward trend in terms of market penetration as more sophisticated applications take over its popularity, usage is still expected to remain high. Comics and other cartoons are expected to make good use of the enhanced graphical capabilities of next-generation messaging technologies such as MMS to pick up on this trend.

SprintPCS has launched The Funniest.com, which is part of Airbourne Entertainment, an SMS and WAP jokes and comic service. Voted the number-one wireless entertainment destination in Canada, The Funniest.com has since formed partnerships with other carriers such as Bell Mobility, Fido (Microcell), and Telus Mobility.

Location-Based Services (LBS)

The business case for LBS is built upon consumers looking for relevance to their situation. Once service/content providers can pinpoint the location of their customers, they can tailor their offers to meet people's immediate needs.

Today's SMS solution is the preferred bearer for the majority of successful LBS, but the imminent arrival of richer messaging technologies such as MMS is set to introduce the ability to increase airtime usage and differentiate their offers to bolster customer loyalty even more. Operators in the United States are currently deploying the infrastructure under regulatory duress (more later), while the rest of the world's wireless players still seek solid proof of commercial viability prior to making the necessary investment.

It is widely acknowledged that location should be viewed as an integral part of a network rather than as an application in itself; the value comes when location is tied to other applications. The most relevant examples of this are in the LBS solutions already commercially available. These tend to be requests for information or chat applications with a localized aspect to their service offerings. The solutions available today are based on cell ID.

Cell ID By identifying the nearest cell to the mobile phone in question, network operators can gain a rough idea where users are. The accuracy of this method varies depending on the cell concentration; it can be accurate to a couple of hundred meters in built-up areas with a large number of base stations or right down to within several miles in rural areas. However, the precision offered by this system is accurate enough for local amenity searches, such as requests for hotels in the area, the nearest Italian restaurant, or the location of a bank's ATM.

The U.K.-based company Mobile Commerce has produced a solution called *Text Hotspots* designed to inform users on the hippest bars, clubs, and restaurants, and give subscribers directions to their nearest celebrity hangouts in London via a simple SMS request. Mobile Commerce combines real-time location data from the mobile phone user's network with the celebrity web site PeopleNews.com's database of city-based celebrity information. Once a user's location is identified, a choice of celebrity venues is offered, and after making a selection, users can opt to receive further details and directions to get to their destination.

The Swedish mobile operator Telia also has a range of LBS applications up and running including a facility for locating friends (called FriendFinder) by linking the location gateway with an *Instant Messaging* (IM) platform. Developed by SignalSoft, FriendFinder enables young people to track whether their friends are in the area and parents can also check up on the whereabouts of their children. It is faster and cheaper to locate 10 friends at one time than it is to call them all individually to determine their whereabouts.

Location-Based Marketing Consumers fear reaching the stage where they become inundated with wireless messages regarding special offers or sales triggered simply by walking down the street with a mobile phone. As with all wireless marketing applications, the emphasis is very much on permission-based campaigns. However, when used correctly, targeted marketing campaigns offering consumers the chance to receive information regarding bargains or promotions while on a shopping trip can be a powerful application.

Case Study: ***Jurong Point Shopping Centre, Singapore***

Who: Mobileway, CELPH, Jurong Point Shopping Centre, and various resident vendors.

Where: Jurong Point, Singapore.

What: Each of the 14 entrances to the shopping center sports a poster and freestanding billboard informing shoppers of the SMS promotion and news service available within the shopping center. It invites shoppers to send an SMS message to the advertised number estimating the number of hours that they expect to spend in the commercial center. After registering, the user will be sent a welcome note giving shoppers the ability to personalize a profile as to areas of interest, such as leisure, food, furniture, clothes, books, and so on. For the duration of their stay within the shopping center, shoppers will receive targeted text messages concerning shop events, promotions, money-off coupons, and incentives for entering a wealth of competitions.

Why: To increase revenue generated per mall visitor.

Outcome: The number of participating vendors grew very quickly following the launch of the campaign to include a whole host of promotions such as discounted books, beauty products, free beverages in onsite restaurants, and competitions to win electrical goods. User statistics still need to be analyzed. Figure 4-6 provides an advertisement for LBS marketing.

911 Caller ID The technology is already available to operators, enabling far greater precision; however, unsure of the potential and commercial viability of the services, they seem reluctant to deploy it. U.S. operators have not been able to take this waiting approach. The *Federal Communications Commission* (FCC) has published a mandate called *Enhanced 911* (E911), which requires that all wireless operators, like their fixed-line counterparts, provide the emergency services with information about the location of 911 callers. Deployment has been split into two phases:

- **Phase I** Cell site ID and callback information must be made available to the *Public Services Answering Points* (PSAPs).

Figure 4-6 CELPH alerts operate at Jurong Point Shopping Centre (Source: Mobileway)

- **Phase II** This step requires that wireless operators provide the PSAPs with more accurate information, which is called *Automatic Location Identification* (ALI).

The deadline for Phase II was pushed back until the end of 2005, but since the tragic events of September 11, 2001, security-conscious Americans have started putting the pressure on for deployment.

Ever in the footsteps of the United States, Europe is developing its own version of the E911 scheme called E112 (112 is the common emergency code shared by European countries). There is still uncertainty as to whether the European Commission will decide to mandate the implementation as the FCC did in the United States or whether operators will be left to roll out the system in line with commercial demand. Ultimately, operators are waiting for a clear sign that LBS will have genuine revenue potential before going ahead with the new technology. Forecasts have not proved compelling so far, with estimates only running at LBS generating at best 3 percent of total operator revenues by 2010.

There are two methods of deploying ALI—network- and handset-based solutions:

- **Network-based solutions** These use base stations to triangulate a device's position, giving accuracy to 1,000 meters on average. The benefit of this approach is that users can enjoy the benefits immediately without having to upgrade their handsets. Voicestream anticipates having a network-based solution in place by the end of summer 2002.

- **Handset-based solutions** These provide a higher accuracy rate to around 100 meters, but this benefit is offset by the need for new terminals to operate the solution. There are two types of handset-based solutions. The first is *Enhanced Observed Time Difference* (E-OTD), which is provided by *Cambridge Positioning Systems* (CPS). Vendors such as Ericsson, Siemens, and Nortel have adopted CPS' technology in their location servers. The second handset-based solution is *Assisted Global Positioning System* (A-GPS), which is offered among others by the Qualcomm subsidiary SnapTrack. In the United States, the trend has been for *Code Division Multiple Access* (CDMA) operators to back A-GPS and for GSM operators to opt for E-OTD. The needs for an embedded GPS chip in the handsets for A-GPS can be expensive and a drain on battery supply.

In-Car Telemetry Telematics, or route guidance, is an appealing application for LBS. Users no longer have to fumble for outdated maps; they simply reach for their cellular phone and obtain clear step-by-step, text-based directions. The most advanced service of this type currently available is provided by a Japanese company called KDDI.

KDDI's Eznavigation service incorporates about 20 different LBS, including route finding. The Eznavigation receives signals from a GPS satellite and measures the user's location, creating only a few meters of error. GPS Map, KDDI's location service for mobile phones, enables the remote tracking of people carrying compatible handsets. It also supports the sending of messages from the monitoring PC to individual mobile phones, enabling dispatched personnel to make speedy responses to instructions issued by supervisors, thus improving operational efficiency. KDDI's service is due to be commercialized in October 2002.

Communication Services

As discussed in the introduction to this chapter, P2P communication remains the main force behind driving messaging traffic and generating revenues. However, additional methods of communication services are available involving third-party service providers that offer text-based content over some form of server architecture. In general, the various forms include the following:

- **Person to person** Voice, e-mail/text messaging, and voicemail
- **Person to machine** Telebanking, financial trading, and electronic shopping
- **Machine to machine** Navigation services and telemetry
- **Machine to person** Automatic alerts and location-based advertising

Extended P2P Communication Many believe that the key to increasing traffic and revenue is in offering enhanced forms of P2P communication services. E-mail is the obvious example and one that is already proving to be compelling in the case of i-mode and the launch of services such as MSN Mobile Hotmail®. MSN now offers its leading Hotmail web-based e-mail service to mobile users over two-way SMS in Europe and Asia. MSN's 110 million Hotmail users on the go have the option to receive e-mail sent by SMS to their MSN Hotmail accounts on their mobile phones. They can reply to those e-mail messages directly to the sender's inbox via SMS and perform other common tasks directly from their phone.

Group messaging is a further extension of P2P communication where individuals can communicate with a group of people via a single message, although the revenue per message is smaller and the traffic levels are higher. The Finland-based company, Popsystems, ran trials at the end of 2001 with a school basketball team. Their results showed that traffic per user increased by 60 percent and revenues increased by around 20 to 30 percent.

Another way to enhance P2P communication services comes in the form of EMS, Smart Messaging, and MMS technologies, where text messages can be enriched by the addition of graphics and animations. EMS and MMS are discussed in more detail in Chapter 8, "Wireless Messaging's Future: A Look Down the Road."

Chat, Flirt, and Adult Dating Services

SMS has had the same effect on chatting, flirting, and dating services as interactive voice recognition technology has had on premium-rate chat lines. Notoriously an area for which consumers are willing to pay, users can enter into communities to converse with like-minded individuals who share common interests. The more risqué field of adult entertainment will come into its own as networks gain speed and allow for the transfer of still and moving images. In the meantime, tamer versions of virtual flirting take place and increase usage. Like the services offered by Link 77 in the United Kingdom, its NataChata service is a highly configurable artificial intelligence chat engine that enables mobile users to engage in fun and provocative chat messaging while maintaining their own anonymity. Location-based technology will also enhance chat services as it offers a method for users to locate other chatters in the same locality.

Instant Messaging (IM) In line with the perception that messaging will remain the killer application for wireless data for many more years, operators are starting to view IM as attractive for new and current subscribers as well as for helping to reduce the subscriber churn that is currently so problematic. IM provides for real-time online conversations and is currently being enabled on various air interfaces.

Although wireless IM might appear to be merely a faster version of SMS or e-mail, the factors driving IM are different due to its conversational quality. IM networks provide information about when selected users are online, so unlike SMS and e-mail, IM users must activate the service and register their presence. This grants users more control over when messages are received, and by streaming text, a person viewing the stream can follow a conversation even if he or she is not providing responses.

The largest drawback to IM over wireless is its intolerance to delivery delays; any pause in the stream can cause serious disruption to the IM conversation. This poses a problem in markets prone to overcapacity where operators draw on every spare bit of bandwidth for voice service and reserve SMS and e-mail for later delivery. This is not acceptable for IM services.

Unified Messaging (UM) Like IM, *Unified Messaging* (UM) is another promising growth area. UM systems provision for a variety of access methods to recover messages of different types, such as text to voice for reading e-mail, text to fax, and the more challenging form of voice to text. Methods to provide UM include a common mailbox consolidating all media; uniform message delivery formatting for each access device, PC, *personal digital assistant* (PDA), or telephone; or a user interface provided by text, speech, or numeric keypad, designed to enable automatic filtering and forwarding according to a user's particular requirements.

The largest challenge to long-term growth in UM is in providing a full host of related services that can build upon the principle of UM, such as extending UM to Unified Communications, enabling the UM interface to act as the entry point to the full communications network including call management, directory services, online billing, and content.

Entertainment Services

Of all the wireless applications designed for personal consumption, wireless entertainment is potentially the most complex area and certainly the one viewed as having the greatest potential. The mobile entertainment industry is made up of two quite distinct industries: entertainment and telecommunications. The former is intent on generating revenues from people through the provision of time killing and fun activities, and the latter has traditionally been involved in raising revenues from cost-efficient communication services. This brings about an interesting marriage with a value chain made up of artists, directors, producers, technologies, and distributors, and complex issues needing to be addressed regarding not only distribution and share of revenues, but also copyright management and royalty fees. The combination of these industries has yet to reach a consensus on the form of the value chain or produce a convincing sustainable business model. Projects under way rely heavily on individual negotiation on revenues and operation. This has obviously had a major effect on the level of content made available for wireless distribution. Branded content is still lacking and the industry still relies heavily on application development.

Ring Tones and Logos Downloadable ring tones and character logos are early examples of mobile entertainment services; these were primarily driven by a desire for personalization rather than entertainment. Ring tones, in particular, were the real driver of i-mode's success and reached enormous popularity levels in Asia and Europe via SMS. Simple pieces of music or animated simple logos downloaded via SMS or the Web to the handset enable users to truly reflect their preference and personality into their phone.

Handset vendors including Nokia, Ericsson, and Siemens recognized the potential in this market and soon developed their devices to include a function for users to design and build their own ring tone directly on the phone or PC, offering the ultimate personalization.

In addition to selling tones and logos over handset manufacturers' web sites, such as Nokia's own Club Nokia site, or directly from *points of sale* (POSs), such as The Phone House, vendors started to use more traditional methods for purchasing downloads, such as premium-rate fixed telephone lines with advertising for such services plastered over newspapers and magazines.

Ring tones will develop in the future and have higher-quality, multitimbral, and polyphonic sounds. Logos will feature richer colors and more animation with the increasing use of screensavers as well as homepage images.

The drawback of ring tones is their effect on copyright. It is estimated that ring tone downloading currently costs the music industry around $1 billion in lost revenue and royalties. In the case of embedded ring tones that come packaged within the mobile handset, tunes are often selected from works for which copyright has expired to avoid the issue. However, downloadable ring tones rely on the integrity of the vendor. Steps are being taken to ensure that this issue is resolved. Nokia has taken one such step. In the summer of 2000, Nokia and EMI initiated a joint venture to provide Club Nokia members the facility to download authentic ring tones from artists featured in EMI's catalog. This marked the first major deal between a record company and a mobile handset manufacturer to combat this problem.

Picture Messaging The picture market is currently one of the largest entertainment sectors. It is made up of four main applications: character downloads/logos, multimedia messaging, postcards, and the adult sector. We already addressed the largest of these four categories—animated logos—in the previous section, where success has been built upon the fact that the low price per image is driving usage figures up.

Multimedia messaging is predicted to be potentially the fastest growing field in mobile entertainment as more sophisticated handsets emerge on the market and bandwidth constraints are relieved with the arrival of messaging technologies such as EMS and MMS. This migration will enable text-based mono formats to evolve to more of a multimedia service, incorporating specialized fonts, colors, images, and audio clips.

The success of this application is dependent upon service providers' ability to maintain the ease of use experienced with the SMS model, and the costs must be controlled to ensure that the content is compelling enough to meet the cost. An early example of this type of picture messaging comes from FunMail. FunMail boasts that it is the world's first system to translate text messages into fun, engaging animation or visuals. Subscribers send a simple SMS text to their friends, which is analyzed in FunMail's server, translated into a relevant cartoon, and added to the message. Users can opt for a message preview option, enabling them to pick and choose from FunMail's cartoon database, which features well-known cartoon series such as South Park and Garfield. The FunMail system can also be easily white-labeled for other content or brand owners. Services such as this attract a great deal of interest from the lucrative youth segment.

A further adaptation to picture messaging is the sending of digital postcards. E-greeting-cards have already experienced great success on the Internet and in spite of the exorbitant cost of text messaging while roaming on foreign networks, users are willing to pay a premium for the convenience of sending a digital postcard. These factors point to a strong case for wireless postcards, particularly with the graphical capabilities heralded by the evolved messaging technologies and *third-generation* (3G) telephony.

Wireless Gaming, Voting, and Quiz Wireless gaming has been a major success in South Korea and Japan, and is widely regarded as a major driver of mobile data usage for mobile networks across the rest of the world. There are a number of reasons for this success:

- There is a demand for time-filling activities for users to indulge in to pass away the minutes while commuting, for example. This places a demand on short-duration games such as quizzes or multilevel arcade-style games where users can conveniently bring play to an end after a short amount of time.
- Processing power and screen quality are rapidly improving in wireless devices. Both of these features are deemed essential for more compelling gaming experiences.
- Demand for platform games as seen with the Sony PlayStation and Microsoft's Xbox remains strong, and the user base for these types of games demographically matches the most active consumers of mobile services.
- Continued growth of Internet parlor games such as Solitaire demonstrate the continued demand for relatively simple games that enable a user to play against others.
- Development in location-based components to wireless technology will bring an additional element to gaming, promoting the rise of gaming communities.
- Downloadable mobile games are already experiencing success in places like Germany. They feature a simplistic business model where users can request extra levels by SMS and get billed accordingly on their monthly bill.

There are a number of key players in this market, but the mobile games market has failed to reach its true potential so far, partly because of the constraints introduced by WAP, making them slow and cumbersome to use. Another reason for this is that big investments are being made up front in development, often by small companies where there are many creative propositions and a great latent demand for mobile games, but not a high willingness to pay.

Java, the open platform introduced by Sun Microsystems in 1991, is expected to wield its influence in the gaming space. Java in a mobile device offers the possibility to easily download updated or new applications into a phone's memory. Since it is both memory and processor intensive, game developers must eagerly await the ubiquitous availability of phones incorporating the slim-line mobile version of Java—*Java 2 Micro Edition* (J2ME).

Simple WAP or SMS games such as quizzes or trivia have been one way to fill the stopgap before more complex gaming environments can be made available. In addition, games such as BattleMail KungFu have converted very well to the mobile medium.

Case Study: *Mobile BattleMail KungFu*

Who: BattleMail, Mobileway, and Siemens.

Where: Worldwide.

What: In December 2001, the mobile version of BattleMail KungFu was launched on the Siemens C45 handset. Players can challenge one another to a virtual KungFu BattleMail, making moves using SMS to try to outwit their opponent with a series of high kicks and karate chops. The objective for the launch of the highly successful multiplayer fighting game was to generate revenue for content providers and mobile operators. This application was the first to follow a revenue-sharing principle. Payment was shared among the participating mobile operators (Siemens, Mobileway, and Battlemail.com).

Why: This revenue-sharing scheme pioneered by Siemens and Mobileway benefits all parties by opening up new channels of revenue for enhanced services across today's incumbent technologies, enabling further differentiation over competitors to ensure retention or growth in market share.

Outcome: A *British Academy of Film and Television Arts* (BAFTA) award nomination in November 2001 and an award for consumer excellence at this year's SMS 2002 conference in London for this mobile version of the game are testaments to the popularity and success that Battlemail.com has achieved for its KungFu feast.

Since the launch, 25 million challenges and more than 20,000 fights have taken place across mobile networks, boundaries, and nationalities.

Interactive TV and Radio Programs The popularity of wireless messaging has prompted broadcast media companies to introduce it into their media mix to drive audience interaction. Recently, there has been an increased inclusion of opinion by SMS and e-mail to current-affair programs, questions into radio broadcasts, and votes of popularity backing participants of game shows.

Voting is a simple but effective way to generate loyalty among television and radio audiences. Users are motivated with the lure of prizes or provoked to place a vote out of loyalty for a favorite pop star or TV game show contestant. Perhaps the most effective example of this has been Endemol's controversial reality game show *Big Brother*. Its format has been rolled out in various forms across the globe. Messaging has been employed as a way to support this format and involve the home viewer in the popularity vote for the individuals undergoing the Big Brother experience. The recent airing of *Big Brother 3* in the United Kingdom brought about record levels of SMS interaction for a television program.

Case Study: Jede Sekunde Ist Wichtig (Every Second Counts)—Interactive TV Program, Germany

Where: Germany.

What: Viewers competed via SMS during an experimental broadcast of Germany's version of the game show *Every Second Counts* to win a prize worth 10 years of free usage of a mobile phone. The goal was to guess the exact number of people in a televised picture. Each SMS guess received the response "Higher" or "Lower" from the SMS-C, enabling the armchair competitors to vote as many times as they liked.

Why: To interact with and increase the loyalty of the channel's viewing audience.

Outcome: Even with no advertising prior to the program, 20 percent of the show's 2.27 million viewers entered.

As well as enticing users to interact by messaging during the airtime of a TV or radio broadcast, media companies have recognized the potential for launching isolated messaging games and quizzes based upon popular game show formats. The most obvious example of this is with the popular show *Who Wants to Be a Millionaire?*

Mobile Music The emergence of standard formats for the digital compression and storage of sound, and music content in particular, is opening up the possibility of downloading music to the wireless device. Music can be downloaded to the device via a fixed-line Internet connection and then played from the phone's memory, it can be downloaded via a mobile Internet connection and listened to while offline, or it can be listened to while online via streaming from an Internet connection.

The capability to stream audio and video over packet-switched networks such as *General Packet Radio Service* (GPRS) and the imminent 3G technologies of *Universal Mobile Telecommunications System* (UMTS) has made mobile music a more viable option. Traditional GSM and CDMA networks proved too slow and costly for effective downloads.

Many issues need to be resolved before music can fulfill its true potential; pricing remains a contentious issue, as does the challenge of copyright management, which is already a problem with phone ring tones. Downloads of entire music tracks pose even more challenges to the management of digital rights and royalties. Many industry groups such as the Mobile Entertainment Forum have launched task forces to address these barriers to deployment.

Marketing Services

The analyst firm Frost & Sullivan predicts response rates to mobile marketing campaigns to be around 40 percent compared with 3 percent from direct mail and 1 percent for Internet banner ads. Although much of this can be attributed to the novelty factor that will inevitably fade, as was the case with fixed-line Internet advertising, there is strong evidence that wireless marketing is a powerful tool. Some of the leading factors contributing to the strength of wireless marketing include cost effectiveness, the appeal to the youth market, and, unlike e-mail, messages have to be opened in order to be deleted, thereby guaranteeing a degree of visibility. Three types of campaigns are open to marketers for engaging consumers using wireless messaging (see Table 4-1):

- **One-off push campaigns** This is the most basic of all campaigns. Marketers can push a single message to a database of mobile users. To ensure that this does not contravene guidelines surrounding spamming, an opt-in by users is essential.

Table 4-1

Comparison of the three types of SMS marketing campaign

	One-off Push	One-off Pull	Continued Dialog
Marketing objectives	Awareness	Promotions	Customer retention
Channel integration	Not needed	Required	Desirable
Channel of interactivity	Low	Medium	High
Database requirement	Required	Not needed	Desirable
Campaign setup time	Days	Weeks	Weeks to months

Source: Forrester Research

- **One-off pull campaigns** This is the next stage up, where traditional marketing channels such as radio or product packaging invite mobile users to respond by messaging, prompting users to automatically opt in. This is more cost effective as the user is charged for the message.
- **Continued dialog** This is the ultimate campaign. SMS interactivity turns into long-term dialog with target customers. An example of this would be if soap opera fans were invited to text their favorite star for advice on certain subjects.

The success of all three of these marketing campaign types relies entirely on an opt-in by the users. Spam is already an issue on the Internet and the wireless industry is pushing for strict guidelines to be drawn up and policed to ensure that wireless spamming does not jeopardize the future of the wireless medium for marketing.

Opt-in Services

Mobile messaging is an ideal tool to get consumers to freely sign up for advertising campaigns and associated services such as newsletters, catalogs, and games. Today marketers have three possible channels for signing up consumers:

- **Postal mail** Usually associated with prize drawings, this can be run through a wide range of supports including the packaging of the product itself. The main drawbacks are that the advertiser has no guarantees on the accuracy of the information retrieved, the sign up is costly to process, and mailing details is not a natural impulse for the consumer.
- **E-mail sign up via a web site** This is both cost effective and impulsive, but it is limited to the Web so it does not enable the advertiser to tease the consumer in a wide range of situations. Besides, many consumers tend to sign up to services with a secondary e-mail address because they often do not want to disclose their professional e-mail for fear of being spammed on a

daily basis. The advertiser also has to deal with consumers mistyping their e-mail details.

- **A call to an IVR system** This is beneficial for campaigns run on a wide range of supports not suited to mail or e-mail sign up such as billboards. Operating an *interactive voice response* (IVR) system is more costly than a web sign up, and the advertiser is entirely reliant upon the information retrieved, risking mistyped numbers and incoherent responses.

Sign Up by Wireless Messaging Recruiting consumers through wireless messaging such as SMS offers many advantages. This can be set up on a large number of supports, particularly billboards. This enables companies to sign up consumers in targeted locations where the consumer would never consider responding to a mailing address. This method is particularly suitable in shopping centers or at the POS of the advertising company, and is both impulsive and reliable as the consumer may not have any information to type as the consumer's mobile number is automatically included in the SMS. It is also extremely cost effective.

Table 4-2

Comparison of sign-up channels available to advertisers

	Mail	Web	Call to an IVR system	SMS
Diversity of support that can be used	***	*	****	****
Impulsive aspect	*	***	***	****
Information retrieved	Mailing address	E-mail	Likely to be a phone number	Mobile number
Cost	*	****	***	***
Reliability of the information retrieved	**	**	**	****

Source: Mobileway

Case Study: *Pepsi "We'll Take You There" Campaign*

Who: Mobileway, Communicator (a Sydney-based interactive marketing agency), and Pepsi.

Where: Sydney, Australia.

What: Pepsi's "We'll take you there" campaign, an SMS and mail-led contest running on the back of Pepsi drink cans, invites consumers to send in their name and a keyword to become eligible for a prize drawing offering winners the chance to attend the Melbourne and Sydney Rumba festivals. The competition was also featured on channel 10's *Pepsi Live* program airing every Saturday evening and Sunday morning.

Why: A trial campaign to investigate the effectiveness of using SMS as a channel for entering promotions.

Outcome: Over 100,000 replies from the combined mail and SMS campaign—21 percent from mail and an outstanding 79 percent from SMS.

- 6,500 responses from the 50-second TV slot on the *Pepsi Live* show, as a result of the rapid response rate that SMS affords.
- Overwhelming evidence regarding the efficacy of SMS as a competition response channel.

Wireless Coupon Services

Wireless coupons, or *m-vouchers*, offer many advantages:

- The consumer is alerted to the arrival of an m-voucher as a new message. Since the message cannot be deleted without being opened, the consumer has to read the m-voucher. This is an advantage over vouchers sent by mail or e-mail, which are often ignored by consumers and end up in the trash unopened.
- The coupon can be received anywhere and anytime granting the advertiser opportunities for intelligent targeting such as sending

a restaurant voucher one hour prior to lunchtime or sending a Direct Marketing TV coupon just ahead of the start of the program. In the case of traditional mail or even e-mail vouchers, the advertiser has little control as to when the consumer actually receives the information.

- Unlike a voice callback, a text voucher such as an e-mail or an SMS is less intrusive.
- The voucher is stored in the cell phone, which consumers always carry, unlike the e-mail stored on a PC or even a paper voucher, both of which are liable to be forgotten by consumers visiting stores to claim promotions.

Redemption of Wireless Coupons or m-Vouchers

By SMS Order Full details of SMS ordering can be found in the "Mobile Billing" section later in this chapter. This is the most instinctive way to use the m-voucher. It is a single-channel process and benefits from the impulsive nature of the SMS order. Figure 4-8 shows an m-voucher being redeemed during an SMS ordering session.

This method has some limitations as SMS ordering is still a very new service for consumers. It may take a long time before this type of service is widely deployed and accepted by a large customer base.

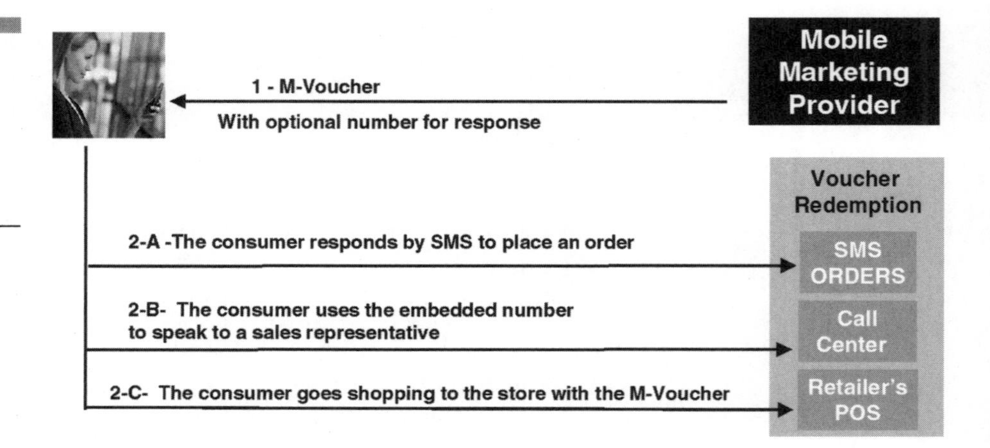

Figure 4-7 Various methods for redeeming m-vouchers (Source: Mobileway)

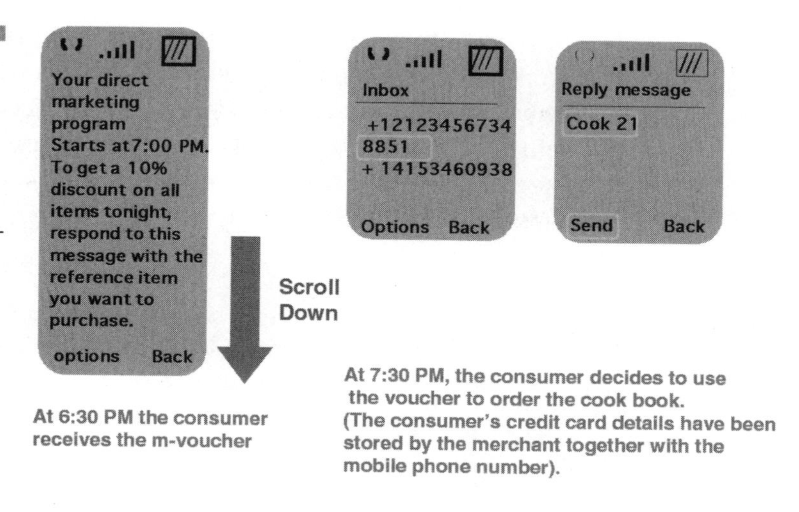

Figure 4-8 Redeeming an m-voucher during an SMS ordering session (Source: Mobileway)

By a Voice Call It may be possible for the consumer to use a phone number embedded in the m-voucher to call the merchant to place the order. This could be a seamless experience if the consumer uses the *use number* option that is featured in most handset messaging menus. However, in most cases, the consumer may feel that switching from the messaging to the voice channel does not offer a very user-friendly experience.

At the Point of Sale (POS) Coupons stored on a user's phone are highly suitable for redemption at the POS in a store as consumers are rarely found to be without their cell phone. However, this poses a technical issue as the POS in a shop is not currently equipped to process an m-voucher in an automated manner. In the future, the Bluetooth protocol could be introduced to interact directly with the POS terminal. This would require the POS terminals to be retrofitted, which is costly in both time and money.

The m-voucher may also be formatted as a barcode that can be processed by standard POS equipment. However, sufficient definition and screen contrast would be needed to make the barcode work; this has been a severe constraint so far. This solution would also require the inclusion of the barcode number, necessitating a mixture of text and graphics in a single message. For m-vouchers that do not require high levels of security, the sales representative may simply ask the consumer to read the coupon number (captured manually on the POS terminal) or even the consumer's mobile number if the POS terminal has access in real time to the database of consumer profiles.

In conclusion, m-vouchers open up a large number of opportunities to merchants. However, the redemption of such m-vouchers is still an obstacle to the large deployment of these services.

Case Study: Dunkin' Donuts

Who: Mobileway, Adreact (mobile marketing solution provider), and Dunkin' Donuts (franchised in Italy by Sweet & Co.).

Where: Rome, Italy.

What: The Dunkin' Donuts campaign in Rome was led by 4 billboards, 2 weeks on the radio, 1,500 leaflets dropped among students, and posters in all 8 of the stores. Donut lovers were invited to enter a prize drawing using SMS. In return, they would receive on their handset a discount or free coffee voucher that could be redeemed in one of the eight Dunkin' Donuts outlets in Rome. Further interaction was encouraged with options to obtain addresses of the franchise outlets, examine statistics regarding Dunkin' Donuts, or inquire about employment opportunities. Users were automatically entered into a drawing for a free Piaggio scooter if they redeemed the SMS coupon in the store and purchased a donut. The campaign is depicted in Figure 4-9.

Why: To increase footfall in the Dunkin' Donuts stores and raise awareness of the brand.

Outcome:

- A 20 percent increase in sales during the campaign period, of which 9 percent can be directly attributed to the SMS campaign.
- Ninety percent of the people claiming the free coffee purchased another product at the same time.

Figure 4-9 Dunkin' Donuts' SMS campaign in Italy (Source: Mobileway)

- More than 50 percent of the respondents opted to continue interaction following the initial SMS response, with 41 percent of those actually requesting the addresses of outlets.

Professional Services

Efficient communication between employees, customers, and suppliers is the lifeblood of an enterprise. P2P wireless messaging makes it possible to reach professionals on the move, whether they are on the factory floor, in the warehouse, visiting a customer, or traveling to the other end of the planet; it is a tremendous efficiency booster.

Keeping organized at work is the second most important thing after communicating. Calendar and address books can be synchronized wirelessly to ensure that everything remains up-to-date.

In addition to relying on people, enterprises rely on information systems to perform numerous vital tasks. Therefore, application-to-person wireless messaging services can enhance the productivity and reactivity of people involved in job functions such as sales, technical support, procurement, and inventory management. All these functions rely upon fast and almost constant access to customer relationship management systems, supply chain management systems, pricing lists, or databases of equipment and parts.

P2P Enterprise Services

E-mail is by far the most dominant means of communicating in the enterprise environment. People often check e-mail first thing in the morning, last thing in the evening, and even during weekends from their home office or portable PC. Although e-mail is less formal than both fax and postal mail, making it unsuitable for binding business transactions, it has a number of benefits such as the capability to handle multicast messages to structured groups and serve as a general data distribution mechanism through setting up distribution lists and allowing file attachments. IM is gaining ground as a faster, less formal method than e-mail, and is only restricted today by its limited security and traceability.

Naturally, extending e-mail and IM to wireless devices is the most obvious application of wireless messaging within the enterprise sector. The clear business rationale is increased efficiency. Urgent messages are handled quicker, less time is wasted driving to rescheduled or relocated meetings, and fewer ideas are lost by waiting until the end-of-day catch-up sessions.

E-mail Although in theory wireless e-mail could be confined to wireless devices, it is most likely to be used as an extension of desktop e-mail operated via PCs, because using yet another e-mail address is the last thing people want. The following are the three ingredients required for an efficient extension of e-mail to wireless devices:

- Always-on data networks to avoid repeated and tedious device-initiated connections to an e-mail server

- End-to-end security between behind-the-firewall enterprise e-mail servers and the wireless devices to ensure privacy over public over-the-air networks
- Large enough screens and usable keyboards on the devices themselves

Research in Motion's (RIM's) Blackberry devices and service have had great success among professionals whose everyday working life is organized around e-mail communications and meetings organized via e-mail. Figure 4-10 shows an example of RIM's Blackberry.

Although heavy-duty users of e-mail need full-blown e-mail clipped to their belt, more casual users can get much of the same benefit in terms of efficiency by only receiving notifications of important e-mail messages. An e-mail notification can be limited to the sender and subject of the message only. This can be suitably displayed on plain cellular phones to avoid the need for dedicated

Figure 4-10 Blackberry from RIM (Source: RIM)

Figure 4-11 XDA mobile GPRS device by O2 (Source: MMO2)

Figure 4-12 Danger Research's e-mail device (Source: Danger Research)

devices. This enables wireless e-mail notification to be extended to any person who goes to work with his or her cellular phone. This method is only suitable for use with nonsensitive notification messages as the lack of end-to-end security when using employees' personal phones can be a barrier to confidentiality. This can be combated by combining the unsecure e-mail notification with a secure browse-back to let users retrieve the body of a message over a WAP session protected by *Wireless Transport Layer Security* (WTLS).

Numerous mainstream IT products can generate e-mail notifications to cellular phones, such as Microsoft's Mobile Outlook and *Mobile Information Server* (MIS), Sun Microsystems' iPlanet servers, IBM's Lotus Notes, and Domino servers. Figure 4-13 depicts an MIS e-mail notification received on an ordinary GSM cellular phone.

Instant Messaging (IM) As with wireless e-mail, wireless IM is most useful when it integrates seamlessly into desktop IM systems rather than being confined to wireless devices only. Wireless IM requires many of the same ingredients as wireless e-mail, such as always-on networks, security, and a suitable user interface on the devices. In addition, IM relies on information concerning the availability of mobile users being communicated to the other participants. Availability should not be confused with *presence*, as the mere fact that a wireless device is on may not mean that its end user is willing to be visible and accessible to fellow IM users.

Figure 4-13 MIS e-mail notification received on an ordinary GSM cellular phone (Source: Microsoft)

Since IM is still somewhat in its infancy, many enterprises tend to use consumer IM systems such as those offered by AOL, Microsoft, or Yahoo! Such systems are readily available for free, and they have all started to feature wireless extensions. IM clients are available from Microsoft and Yahoo! for Palm and PocketPC devices. AOL has also integrated an IM client into selected cellular phones. However, the absence of enterprise-grade security means that public IM systems will be of limited use in professional environments.

General-purpose enterprise collaborative tools from IBM/Lotus, eRoom Technology, PlaceWare, and WebEx have built-in IM features that can be extended to wireless devices using the appropriate client software. Dedicated enterprise wireless IM offerings are also available from companies like Openwave or Ecrio. Figures 4-14 through 4-16 provide graffed examples of how an IM session will appear on a typical wireless phone.

Personal Information Management (PIM) *Personal Information Management* (PIM) systems have become wireless through the ever-increasing use of Palm OS and PocketPC PDA devices coupled with the widespread use of Microsoft's Outlook with which such devices can synchronize. Synchronization via wireless messaging is available from companies like Pumatech and Fusion One. Nokia's Smart Messaging phones can receive vCards as encoded short messages and store them as new entries in the phone's address book.

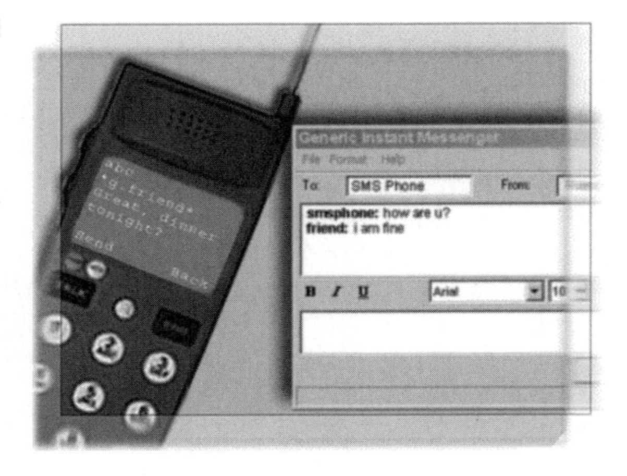

Figure 4-14 Private Web-to-phone IM system from Ecrio (Source: Ecrio)

Figure 4-15 Openwave's IM client for cell phones (Source: Openwave)

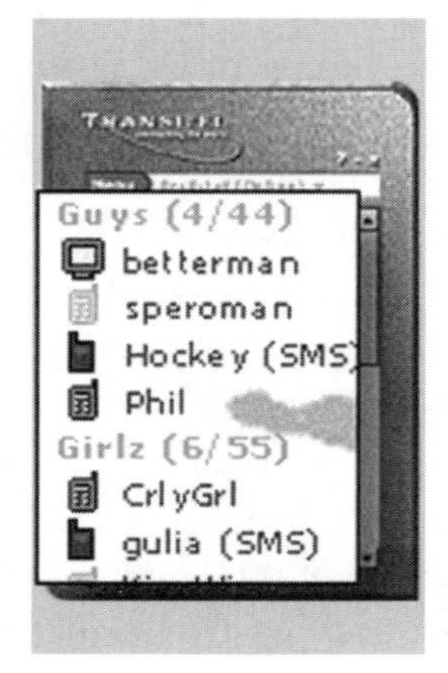

Figure 4-16 Openwave's IM client for desktops (Source: Openwave)

Companies such as Good provide dedicated wireless devices that have synchronized calendars, contacts, and e-mail facilities. Figure 4-17 provides an example of a technology synchronized messaging device by Good. Figure 4-18 shows an example of an MIS Outlook calendar entry reminder.

Application-to-Person Enterprise Services

In recent years, enterprise information systems from mainstream suppliers such as BEA, IBM, Microsoft, Oracle, PeopleSoft, SAP,

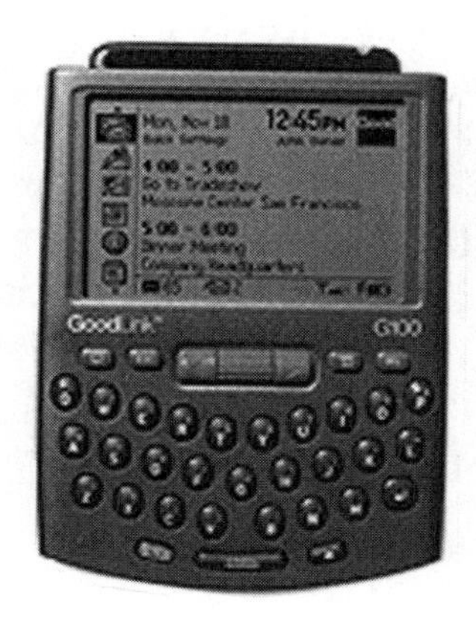

Figure 4-17 Technology synchronized messaging device by Good (Source: Good)

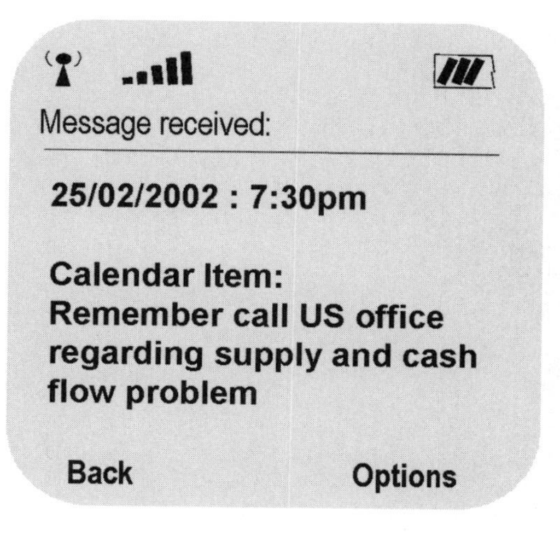

Figure 4-18 MIS Outlook calendar entry reminder received on an ordinary GSM cellular phone (Source: Microsoft)

Siebel, and Sun Microsystems have been released with wireless extensions, enabling remote access from cellular phones and PDA devices. Applications have ranged from generic services such as remote access to databases to more specialized functions relating to *Customer Relationship Management* (CRM), *Enterprise Resource Planning* (ERP), and *Sales Force Automation* (SFA). Smaller, more specialized suppliers have also developed products of their own or products that complement those of the larger vendors. Figure 4-19 is an illustration of Siebel's CRM application.

The rationale is clear: Give mobile professionals anywhere access to the resources they need to conduct business in the field without running the risk of unsynchronized, out-of-date, or duplicated data.

With the advent of web services where data and processing resources become distributed over *wide area networks* (WANs) rather than centralized at the enterprise's premises, remote access becomes a fact of life whether users are mobile or not. Consequently, remote mobile access through wireless messaging is about to become an intrinsic part of new-generation IT products rather than an after-market add-on.

eXtended Markup Language (XML) will play a major role in new-generation application-to-person wireless messaging as the technol-

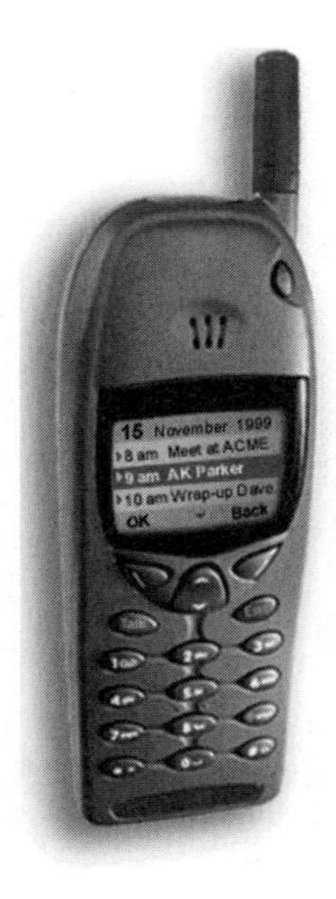

Figure 4-19
Siebel sales wireless screen shot (Source: Siebel)

ogy is ideally suited for future-proof interfacing with applications as well as adapting the presentation of information to different devices and user interfaces.

Notifications from Enterprise Applications In addition to enabling a mobile device to access enterprise data remotely, wireless messaging can be used to alert mobile users to an enterprise application event deserving their attention. By pushing alerts that do not follow any regular time pattern, wireless services can be extended beyond those applications that mobile users will customarily log into remotely. Events generating an application alert can be very diverse, such as a change in pricing, a supply running out, or a shipment going through.

It is likely that wireless alerts will be increasingly used to invite a user to log back into a particular application from a WiFi (high-bandwidth) hotspot as available in offices, hotels, airports, or coffee houses. Low-bandwidth always-on wireless messages reaching even the simplest of cellular phones can be the first step in a much more sophisticated wireless transaction carried out from a high-bandwidth hotspot via a laptop computer.

Summary of Professional Services Professional applications of wireless messaging are still in their infancy, but this is one of the most promising fields. It has a very clear business rationale, no technical obstacles, and strong drive from the mainstream IT vendors.

Although wireless messaging currently serves as a bridge between standard enterprise platforms and wireless devices, architectures will change significantly in the near future with the advent of web services, XML, and WiFi networks.

Wireless Payment and Mobile Commerce (M-Commerce) Services

Recognizing the enormous revenue-earning potential in carrying out secure payments directly from a cellular phone has ensured that building a secure model for *mobile commerce* (m-commerce) has become a priority among network operators, merchants, and financial institutions alike.

This chapter differentiates two types of services:

- **Micropayments** These are low-value payments not usually processed by debit/credit card issuers. These payments typically refer to goods and services under $10. The $10 cutoff is an approximate value corresponding to today's credit card fee structure; the processing and transaction charges on credit card purchases offer little, if any, economic benefit for goods and services below this value.
- **Macropayments** These are higher-value payments that can be charged to a debit/credit card account. In this case, the wireless device may be used to place the order or authenticate with the issuing bank.

Using the Wireless Device for Microbilling Services

Various Billing Mechanisms Available With the development of both web and wireless services, many merchants and content providers have entered the business of delivering virtual goods and services. One critical aspect that needs to be addressed is whether they have the capability to charge the consumers for these goods. Because the amount to be charged cannot usually be processed by a debit/credit card, three major solutions are emerging:

- *Stored value accounts* (SVAs)
- Premium SMS
- Mobile billing

Stored Value Accounts (SVAs) The SVA is usually activated or reloaded by the consumer using a debit/credit card. This payment mechanism is widely used by many web merchants when a login and password can be prompted following a URL redirection to the SVA provider. This system is much less convenient when the ordering channel is not via the Internet.

The main drawback is that the consumer may never get to spend the whole amount that has been reloaded, particularly since the SVA providers are not interoperable. Whereas an SVA provider usually offers the service across multiple merchants, the large number of SVAs currently deployed makes it almost impossible for the consumer to use a single SVA as this would result in having to activate and load a new SVA with different merchants.

Premium SMS This has been the most popular payment mechanism to date among content providers in Europe and Asia. It can be used in two ways:

- **An SMS *mobile originated* (MO)** The consumer places the order by SMS. The SMS is sent to a short number dedicated to the service, which triggers a premium billing for that SMS. The request and billing are done in one step through the premium SMS MO.
- **An SMS *mobile terminated* (MT)** The consumer signs up for the service and agrees to receive the goods or alerts at a premium rate. The billing and delivery of the service are done in one small step through the premium SMS MT. For instance, this is what MSN uses in Asia for its e-mail alert service Mobile MSN Hotmail. Each alert is delivered at a premium rate.

Mobile Billing Using the mobile operator phone bill offers more flexibility to the consumer since he or she does not have to prepay an account, which may not be compatible with other merchants. However, the large number of mobile operators and the very diverse types of billing systems that each of them is currently using is an obstacle for content providers to operate such a service themselves. Some service providers (SMS aggregators or payment gateways) are emerging in order to provide content providers with a single access for billing across multiple operators in a seamless manner. The payment flow may involve one or two stages:

- **One stage** A billing order is sent prior to delivery of the good. Problems may arise with this system if the good cannot be delivered and the account of the consumer cannot be recredited.

- **Two stages** An authorization request is sent before delivery of the good and a billing order is sent following delivery of the good. This approach is ideal for mobile operators who cannot recredit the subscriber's account if delivery does not occur for whatever reason. The drawback with this scenario is if the billing order fails to be generated on delivery of the good. This is a danger particularly with prepaid airtime accounts where users may have exhausted their phone credit.

In both cases, if the order is not placed by SMS, a separate cardholder authentication may be required before an authorization request is sent to the mobile operator.

Microbilling Scenarios To summarize, the microbilling service can be split into three steps:

- Ordering channel (via the Internet, SMS, or Mobile IP)
- Billing channel (a premium-rate SMS message, a billing order to the mobile operator, or an Internet-based SVA)
- Delivery channel used for the service or good (SMS, Mobile IP, or the Internet)

Table 4-3 lists various billing options available according to the ordering channel and the delivery channel.

Issues That Need to Be Addressed In micropayment, as with all forms of mobile payment, various issues need to be addressed:

- The consumer disputing the transaction
- The need to recredit the consumer account
- The case where the consumer cannot be billed
- The effect on convenience for the users when the ordering channel and the delivery channel are not the same

Dispute of the Transaction In the case of a dispute over microbilling, regardless of the billing channel used, the payment service provider is most likely to recredit the account without any further investigation. Any additional cost related to the dispute would certainly be far too high compared with the actual amount of the transaction.

Table 4-3

The various billing options available according to the ordering channel and the delivery channel

	Paper Catalog + Telephone Order (TO)	Paper Catalog + SMS Order	WAP Shopping and Order	Web Shopping and Order
Immediacy of access to the shopping media	****	****	*	*
Rapidity of shopping	**	**	*	***
Richness of content	***	***	*	****
Immediacy of the order	**	****	***	***
Availability of the technology	****	***	**	**

Source: Mobileway

Recredit After the Transaction Has Been Billed Following billing, the payment service provider may have to recredit the account if the consumer disputes one of the following: the transaction itself, the delivery of the good, or the quality of the good. If billing is via an SVA, recrediting a customer is straightforward. However, many mobile network operators are not able to recredit an account if the subscriber has a prepaid or postpaid billing relationship.

The Consumer Cannot Be Billed In the case of an SVA, this is not an issue; the account can always be debited from a prepaid value. The SVA provider may face disputes related to the reload or top-up transaction, but this is an issue related more to the payment guarantee of macropayment transactions.

In the case of a mobile operator billing, there are various scenarios:

- This is not an issue in the case of one-stage billing or if the consumer is a prepaid subscriber and the billing order is sent prior to delivery of the good or service.

- This is an issue if the billing order is sent following delivery of the good and the prepaid airtime account has insufficient funds for the balance, or if the consumer is a postpaid subscriber and does not pay his or her phone bill. In both cases, the mobile operator or the content provider will be liable for this. This depends on the contractual relationship in place between the mobile operator and the payment service provider.

User Convenience Combining different channels (Internet/mobile) for ordering and delivery may alter the user experience. In the case of an Internet service, the consumer may prefer to place the order through the Internet and use a web-based payment mechanism such as an SVA. In the case of a mobile service, a premium SMS service may be the more appropriate channel. However, mixing both channels could be considered too. A web browser is indeed a pull mechanism that may be well suited for placing an order (although it is not as immediate as sending an SMS). For instance, today many ring tones are ordered from web portals by using an SVA.

Tables 4-4 summarizes the benefits of each payment mechanism for either Internet or mobile services.

Using the Mobile Device for Macrobilling and M-Commerce

This section focuses on the opportunities associated with using a wireless handset to conduct a macropayment transaction using a debit/credit card. The device can be used as an enabler during the

Table 4-4

The benefits for Internet and mobile services

	SVA	Premium SMS	Mobile Billing
Re-credit the account	***	*	*
Dispute	*	***	*
Consumer Billing does not happen	****	**	*
User convenience	****	*	**

Source: Mobileway

shopping phase or as a means to authenticate the cardholder. The objective is to determine which transactions would benefit from using a wireless device and at which stage in the transaction the device should be used.

Transactions Where the Wireless Device May Add Value Macropayment transactions can be split up into three main categories:

- POS transactions
- *Mail order/telephone order* (MO/TO) transactions
- Internet-based transactions

The value that the new technology brings addresses fraud reduction, a better consumer perception that results in higher transaction volumes (consumer perception is often associated with security, but it is sometimes subjective and unrelated to an actual reduction in fraud), and finally, a faster transaction for the consumer.

POS Transactions Today's POS purchases involve a two-factor authentication via electronic terminals accepting plastic cards. The two factors are a card plus *personal identification number* (PIN) or signature followed by an online authorization. This guarantees low fraud, fast transaction processing, and a large consumer adoption.

In this case, the value of using a cellular phone at the POS is hard to identify, although there is a window of opportunity in those countries where the penetration rate of cellular phones is higher than that of credit cards. There is the case during physical transactions where the merchant is not equipped with an electronic terminal, such as market stalls and taxis.

Mail Orders/Telephone Orders (MOs/TOs) Cellular phones assist existing TO merchants in the following ways:

- Increase the immediacy of the order by using instantaneous channels such as SMS messaging.
- Improve consumer perception of security by providing a mobile messaging authentication service.
- Reduce fraud for certain merchants.

Internet-Based Purchases For Internet orders, the consumer perception of security and the actual fraud level could be improved by the introduction of cellular phones. However, a major obstacle may be consumer convenience. Many services today offer cardholder authentication through a password in a *Secure Socket Layer* (SSL) session. This may be quicker and easier for consumers than using their cellular phone.

The Remote Macropayment Value Chain The value chain of the remote shopping transaction is as follows (see Figure 4-20):

- The purchase is initiated by the consumer based on a wide range of shopping media.
- The consumer places the order.
- The merchant uses various systems available to verify that this is indeed the cardholder making the transaction.
- The merchant requests authorization for the payment from the acquirer.

The Shopping Media Various remote shopping media are currently available to consumers (ranked according to transaction volume):

- **Paper catalogs** This is probably the most widely used media for remote shopping. It is immediately accessible to the consumer and is usually associated with a special enjoyment.

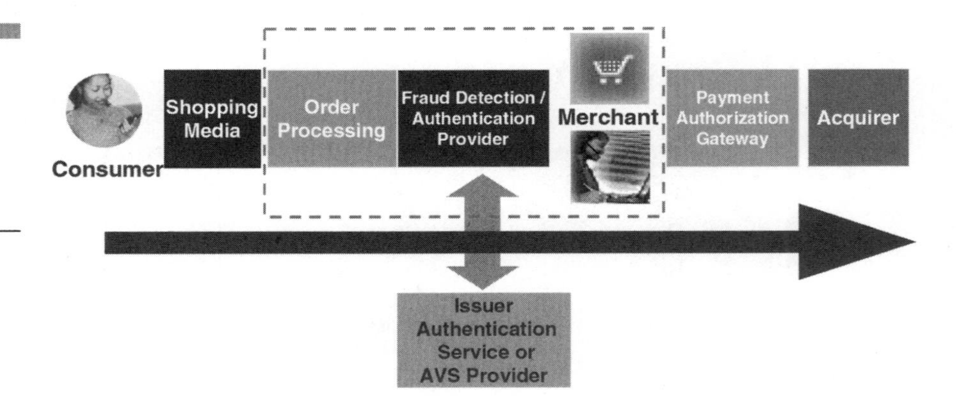

Figure 4-20 Remote macropayment value chain (Source: Mobileway)

- **Voice conversation with a sales representative** Although calling a merchant is often associated with long waiting music, this media remains very popular for any item that can be purchased without a catalog.
- **Direct marketing TV programs** This method is quite popular and reasonably effective.
- **PC web shopping** Even in the United States, 92 percent of households rely on dial-up Internet connections. The immediacy of the shopping experience is usually not as high as for a catalog. On the other hand, this media is best suited for purchases requiring a search engine (travel or books not listed as bestsellers by book clubs) or purchases where the consumer may register a standard basket (such as groceries).
- **WAP shopping** So far this medium has not been adopted by consumers for commerce applications. Opportunities for WAP shopping still need to be determined.

Order Processing Once the consumer has decided the good he or she wants to buy, various ordering channels are available to the consumer:

- During an Internet order, the consumer discloses his or her credit card details via an SSL session (if the shopping took place through WAP, a similar process would take place during a WTLS session).
- TOs are confirmed with a credit card number.
- MOs rely on a check or occasionally a credit card number written on the order slip.

Figure 4-21 illustrates how items were paid for in the United States.

The Cardholder Verification Before requesting authorization from the acquirer, the merchant may verify that this is indeed the cardholder making the transaction.

This is usually done through an *Address Verification System* (AVS) where the consumer can request service providers to verify the billing address that the consumer gave to the merchant. Even

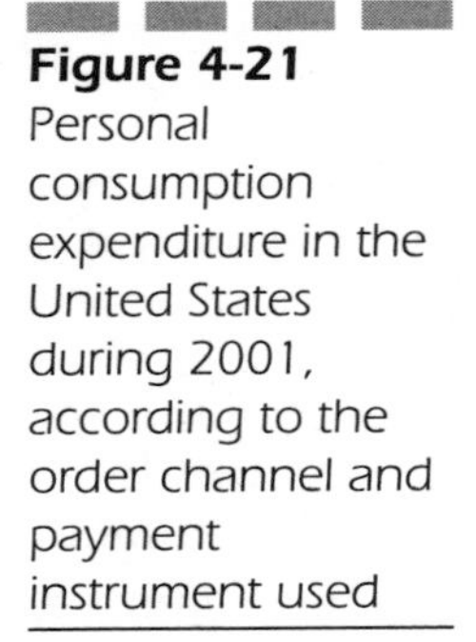

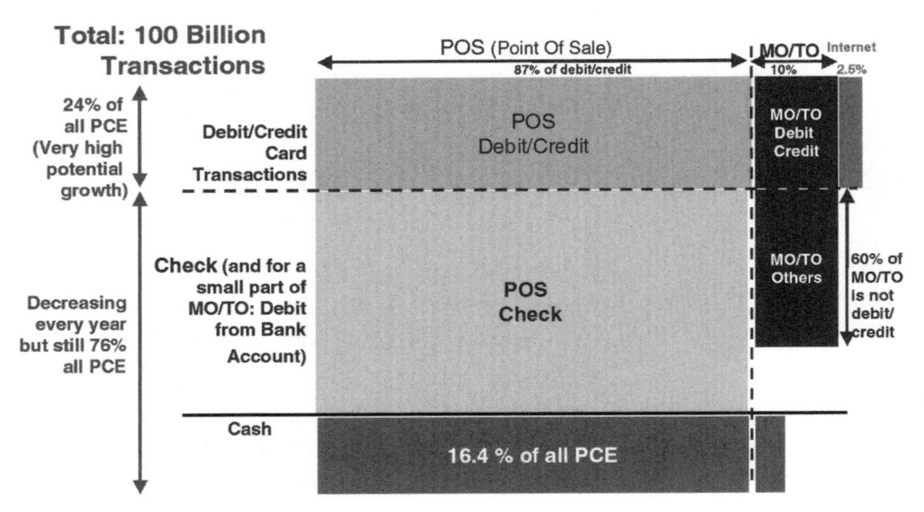

Figure 4-21 Personal consumption expenditure in the United States during 2001, according to the order channel and payment instrument used

though this is a common practice in the United States, the AVS is prohibited in other countries like France in order to protect the privacy of cardholders. New services are currently emerging in order to request the authentication of the cardholder directly using different channels and devices.

Opportunities for Wireless Services in Remote Purchases

Three opportunities have been identified for wireless services related to a remote purchase in which payment is processed using a debit/credit card:

- An ordering service using SMS
- A wireless messaging cardholder authentication service
- A wireless messaging transaction alerting service

Ordering by SMS

So far, cellular phones have had limited success as shopping tools. The main pilots undertaken to date all used WAP browsers and experienced the following limitations:

- The shopping experience is slow when compared to browsing on the Web or through a paper catalog.
- The content available for goods to be purchased in this way is limited.
- While on the move, consumers still prefer to shop by voice, and web sites or paper catalogs offer a much wider range of information.

The most promising application for wireless shopping identified so far is for ticketing services; consumers can purchase cinema or theater tickets using a text message or WAP.

Wireless Messaging as a Simple Ordering Tool

Whereas mobile phones may not be ideal tools for shopping, they may be used as a simple mechanism to place the order as a complement to existing shopping media. Indeed, it helps to address the drawbacks encountered with current mail or telephone ordering channels, including

- *The merchant experiences a high cost.* Approximately $5 per TO and around $2 for an MO.
- *The ordering process can prove tiresome for the consumer.* MOs can be time consuming, which prevents impulse buying, resulting in lost sales opportunities for the merchant. For TOs, the consumer often is subjected to long periods of hold music and IVR systems before finally placing an order with a sales representative. Additionally, caller ID is an insufficient method

Figure 4-22 SMS CheckOut—ordering via SMS (Source: Mobileway)

The consumer must have previously registered his/her credit card details together with his/her phone number with the catalog publisher or TV shopping channel:

1 – The Consumer sees an item he/she wants to buy in the TV program or the paper catalogue of this merchant.
2 – The consumer types the item reference in an SMS and sends it to the merchant service number displayed on TV/catalogue.
3 – The SO module processes the order by parsing the SMS and identifies the consumer with the originating phone number of the SMS.

to verify the consumer's identification so another step is added into the phone order. For many items that do not require any phone conversation (such as CDs, books, and so on), the consumer would prefer a faster way to place the order. Figure 4-22 provides an example of an order via SMS.

In order to overcome these issues, part of MOs or TOs could be replaced by a wireless messaging order.

This new ordering channel is ideal for the following situations:

- Purchases that are not required in any large quantity (such as books, DVDs, and so on)
- Purchases that do not require any conversation with a sales representative
- Purchases that are made or repeated on a regular basis—for example, contact lenses

Why SMS Is Better Suited Than Other Mobile Channels?

SMS is better suited for purchasing because:

- No WAP or GPRS phone is required as SMS is available on 1 billion cellular phones worldwide.

- There is no URL to bookmark (instant order through the short code).
- No browsing is required.
- No login is required (the phone number included in the SMS header makes the transaction *device present*).
- The consumer always keeps a record of the transaction.
- Consumer adoption is a testament to the popularity of SMS, with 1.3 billion SMS sent each day.

Since the consumer only texts a reference of the item, the value of using more advanced mobile messaging mechanisms such as EMS or MMS is rather limited. Table 4-5 provides further detail into various ordering channels.

SMS CheckOut provides the consumer with the most immediate and impulsive channel available. The consumer just needs to text the reference of the item to the short service number of the merchant and the order is placed.

Benefits of SMS CheckOut to Consumers SMS CheckOut is perceived as more secure because no credit card number has to be disclosed

Table 4-5

Comparison of the various ordering channels

	Paper Catalog + Telephone Order (TO)	Paper Catalog + SMS Order	WAP Shopping and Order	Web Shopping and Order
Immediacy of access to the shopping media	****	****	*	*
Rapidity of shopping	**	**	*	***
Richness of content	***	***	*	****
Immediacy of the order	**	****	***	***
Availability of the technology	****	***	**	**

Source: Mobileway

and no one can place an order with this merchant without the consumer's mobile phone. It is also the fastest way to place an order. The benefits to consumers of SMS CheckOut appear in Figure 4-23.

Benefits to Merchants For merchants, the benefits of SMS CheckOut are the lower transaction costs than equivalent MOs or TOs. The benefits to merchants of SMS CheckOut are shown in Figure 4-24.

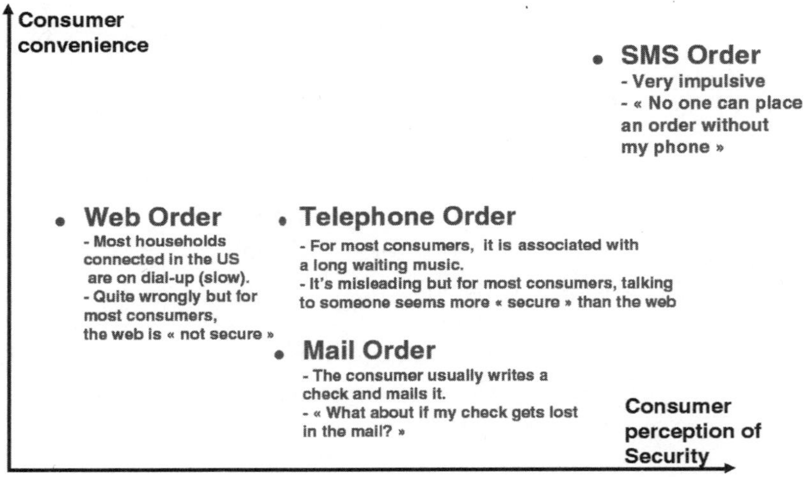

Figure 4-23 Benefits to consumers of SMS CheckOut (Source: Mobileway)

Consumer convenience

- SMS Order
  - Very impulsive
  - « No one can place an order without my phone »
- Web Order
  - Most households connected in the US are on dial-up (slow).
  - Quite wrongly but for most consumers, the web is « not secure »
- Telephone Order
  - For most consumers, it is associated with a long waiting music.
  - It's misleading but for most consumers, talking to someone seems more « secure » than the web
- Mail Order
  - The consumer usually writes a check and mails it.
  - « What about if my check gets lost in the mail? »

Consumer perception of Security

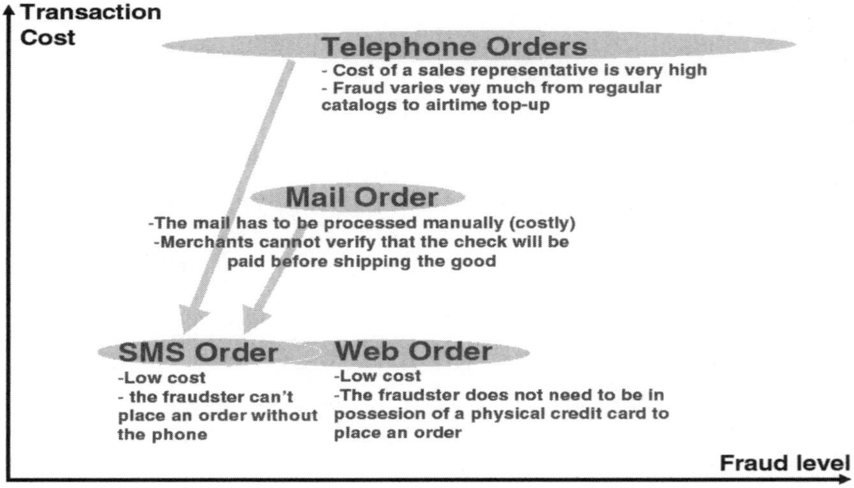

Figure 4-24 Benefits to merchants of SMS CheckOut (Source: Mobileway)

Transaction Cost

Telephone Orders
- Cost of a sales representative is very high
- Fraud varies vey much from regaular catalogs to airtime top-up

Mail Order
-The mail has to be processed manually (costly)
-Merchants cannot verify that the check will be paid before shipping the good

SMS Order
-Low cost
- the fraudster can't place an order without the phone

Web Order
-Low cost
-The fraudster does not need to be in possesion of a physical credit card to place an order

Fraud level

Wireless Messaging as an Authentication Mechanism for Debit/Credit Card Payment

Wireless authentication can be used to secure debit/credit card payments when the order is placed by SMS or a voice call. The main benefits for an issuing bank are the reduction in fraud and the improvement in consumer perception of security and in creating a new innovative service to boost new cardholder acquisition.

Three main channels are available for this type of authentication:

- Wireless messaging such as SMS, EMS, or MMS
- *Unstructured Supplementary Services Data* (USSD) in the push version
- Voice callback mechanisms

A brief comparison between these channels shows that authentication sessions based on USSD or voice will fall out if the consumer cannot make the transaction in seconds, whereas an SMS prompt is automatically stored in the phone, giving the consumer the chance to respond to it at a later stage at his or her convenience. Unlike voice, with SMS the consumer can always verify the origin of the message easily by the short code used to originate the message (see Figure 4-25).

Unlike USSD, SMS is available on any GSM phone. At the end of 2001, only 65 percent of handsets supported USSD push. Furthermore, USSD is not a standard service; it is only available from

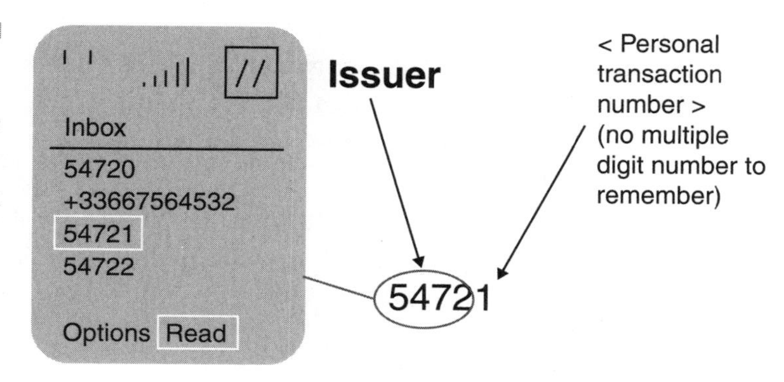

Figure 4-25 Verify the origin of the message by the short code (Source: Mobileway)

wireless network operators who have implemented it within their network. To date, only a few mobile operators have implemented such a service. Also, the consumer adoption of SMS has been overwhelming, whereas the possible usage and understanding of USSD by consumers is unknown, as it cannot currently be used as a P2P service. Table 4-6 provides a comparison of the various authentication methods.

A Wireless Messaging Authentication Service Two main technologies can be used for authentication using wireless messaging:

- **A client-based approach** For example, this approach can involve a *SIM Toolkit* (STK) applications held within the handset, which can be downloaded onto the *Subscriber Identity Module* (SIM) card. This solution offers robust security with end-to-end encryption of the data sent over the air. Around 25 mobile operators have launched such a technology. However, most of the services have not progressed farther than the pilot phase and the applications implemented are proprietary to each mobile operator.
- **A server-based approach** This enables any service to run on any *second-generation* (2G) cellular phone without any special client. This service is not as robust in terms of security as the STK solution, but it enables a wide reach to potentially 1 billion mobile consumers with a fully interoperable and seamless service across multiple mobile operators.

The *Mobile Transaction Tracker* (MTT) solution from Mobileway appears to be emerging as a standard for such services. The MTT global interface can be downloaded from the Mobile Payment Forum's web site at www.mobilepaymentforum.org. The mobile authentication of the cardholder takes place prior to the merchant's request for payment authorization.

Utility Bill Payment Although the majority of bill payments are still processed by mail, utility companies are making great efforts to

Table 4-6 Comparison of the various authentication methods

	Voice Callback with IVR System		Mobile Messaging Password		USSD	
Availability of the technology to consumers	No client needed	****	No client needed	****	Most networks and some handsets do not support USSD push.	*
Origin of the payment offer	Caller ID is not always reliable for voice calls.	*	Service number	***	Service number	***
Network reliability	Voice session can be interrupted.	**	Store and forward	****	Session based	**
Trace of the transaction	No record	*	The payment offer is stored in the FM inbox.	****	No record	*
User convenience	A voice callback is intrusive. Need to type PIN while listening to voice instructions	*	The session can be postponed (store option).	****	The session cannot be postponed.	**

Source: Mobileway

reduce the cost of processing these costly payments. The following alternative systems have emerged:

- **Automatic debit** Many consumers opt for an automatic debit service from their checking account in order to settle their bill. However, many consumers are still reluctant to activate a service over which they feel they have no control.
- **Debit orders** Some utility companies have put in place billing orders where the consumer simply needs to detach it from the bill, sign it, and mail it. In this scenario, the consumer has previously registered his or her bank account number with the utility. This debit order mechanism is used in many European countries. Even though this method is cheaper to process (utility account number and bank account details are on the same paper, enabling optical process), the utility company still has to process the mail and the paper billing order for presentation to the issuing bank.
- **SMS debit order** This method enables the cost associated with the debit order to be lowered by using SMS as a mechanism to confirm the payment of the bill. Where possible, the consumer receives an SMS reminder with a summary of the bill and can choose to respond to it with his or her password to trigger the payment (see Figure 4-26).

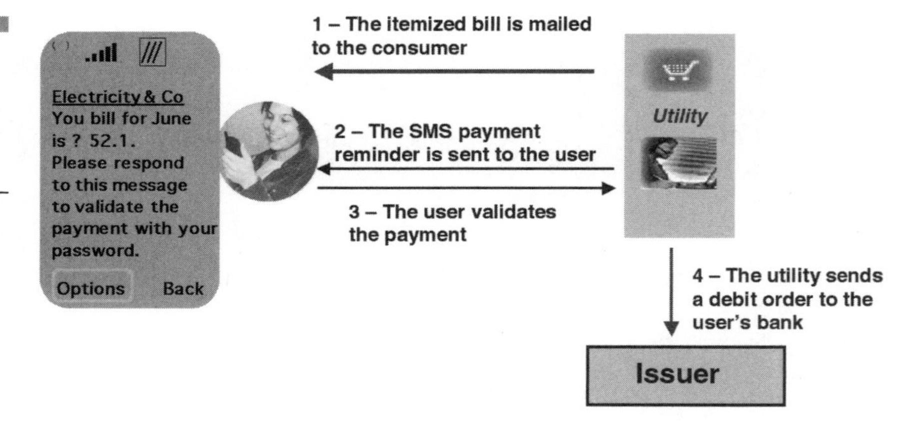

Figure 4-26 SMS reminders for utility bill payment (Source: Mobileway)

This SMS trigger system has the following benefits:

- The system is even more cost effective for the utility company than a paper debit order.
- For the consumer, it is quicker and cheaper than sending a debit order by mail as an SMS is cheaper than a domestic stamp.

Notification Services to Improve Fraud Detection Systems

Each time a transaction is deemed potentially risky, the cardholder is informed by a notification message. This could include alerts to unusual purchasing behavior by the consumer, a transaction of high value, or a transaction taking place abroad on the same day as one carried out domestically. If the suspicion is correct and the consumer did not carry out the transaction, he or she can contact the bank via a voice call or directly by responding "no" to the mobile message. Requiring no merchant upgrade, this notification service may prove to be an excellent alternative to voice alerts for many issuing banks for both remote and POS payments.

This system, however, cannot prevent a fraud on the transaction that is notified. Indeed, the timeout on the authorization response sent by the issuer is too short to ping the consumer on his or her mobile phone. However, this is a useful service to prevent any additional fraudulent transaction, which is particularly pertinent given that fraudulent transactions are usually multiple.

The Future of Wireless Messaging Applications

A number of factors have been set to have the greatest influence on the shape of future wireless applications. These include

- Improvements to the user interface on handsets
- The addition of new handset functionality

- The emergence of other nomadic products capable of communicating wirelessly either by themselves or in cooperation with handsets
- The availability and speed of networks
- The dispersal of services into new human cultures and geographies

User Interface Improvements

Significant improvements to three elements affecting the user interface will give rise to new wireless applications:

- **Screens** The natural and expected evolution of *liquid crystal displays* (LCDs) to higher resolution and brighter colors at reduced power consumptions will increase the suitability of handsets for games and picture-sharing applications. More radical technology changes are on the horizon to enable entirely new types of displays with much improved electromechanical properties, such as screens at near-zero power consumption and those with the capability to wrap screens around edges. This could lead to the appearance of devices with small auxiliary screens for displaying functions such as notification messages, for example. However, screen sizes on handsets or PDAs are unlikely to increase significantly because the devices themselves will continue to be small and portable. Virtual screens generating a much larger floating image in front of the user's eyes are unlikely to become mainstream as they require users to wear some kind of eye device.
- **Keyboards** Thumb keyboards introduced with RIM's Blackberry device will appear in a larger number of PDAs and combination phones/PDA devices. This is set to further increase wireless e-mail and texting usage. However, only a small percentage of users are comfortable with such keyboards, as they require both practice and dexterity. Up-and-coming virtual keyboards that get projected onto any flat surface in front of the device can offer the full-fledged keyboard and mouse functionality found on laptop PCs. Virtual keyboards are

possibly the one ingredient that will best close the gap between the world of cell phones/PDAs and the world of PCs. This would be an enormous boost to wireless Internet applications.

- **Voice input** In many circumstances, while on the move, screens and keyboards remain impractical or even dangerously distracting. In these cases, voice remains the obvious way to communicate both with other people and other applications or systems. Voice IM or *push to talk* is where phones double as walkie-talkies; Nextel in the United States was the first to introduce this. With the improvements of always-on data networks such as GPRS, *Enhanced Date Rate for Global Evolution* (EDGE), 1x-RTT, and their higher-speed 3G successors plus technologies such as the *Session Initiation Protocol* (SIP), voice IM can become widespread by transporting snippets of voice over data packets rather than by making phone calls. Voice IM could become a sibling of SMS, where users could always be given the choice of typing versus speaking and reading versus listening. Voice IM over data packets can also be used for group calls or for interfacing with voice-based services such as TellMe and BeVocal.

Device Features

Typically, combination devices do not fare well in the consumer electronics industry. Even when combining features seems logical such as TV + VCR or printer + fax, these combinations end up representing a very small percentage of the market compared to separate dedicated devices. Like cellular phones, people buy them to communicate, and although it makes sense to combine a cell phone with noncommunication functions such as an MP3 digital music player, it is doubtful whether phone/MP3 players will ever extend beyond marginal distribution.

Address books and appointment calendars are strongly linked to communication functions so they seem like logical add-ons inside a phone. The synchronization of address books and calendars is an obvious application for wireless messaging, especially for business users. In the same manner, low-end digital cameras integrated with phones are an attractive proposition as casual pictures are normally taken to share with other people. Photo swapping among friends and family is likely to become a very large part of P2P MMS services.

Since the mobile phone is a personal device that is not usually shared with anyone including close family, it is an ideal place to handle other personal information that needs to be carried around and communicated from time to time to other parties such as banking cards, coupons, tickets, and so on. Consequently, security surrounding such personal credentials must be tight. New generations of secure chip technology used in SIM cards and their successors *Universal Subscriber Identity Module* (USIM) and *WAP Identity Module* (WIM) cards can be loaded over-the-air with transactional Java applets providing banking-grade protection. This can also be combined with appropriate secure messaging and secure backend servers for end-to-end transaction integrity, authenticity, and privacy.

Emerging Nomadic Products

The list of existing and future nomadic products to foster new wireless messaging applications is long and growing by the day. Some of these products include

- **Motorcars** The most obvious nomadic product is probably the car. The privacy of cars, combined with the time lost in traffic, makes them an ideal place from which to communicate. Aside from the obvious integration of cell phones into the dashboard of automobiles, it is expected that car-optimized wireless communication devices will be adopted as the appropriate user-interface for drivers. For example, this could include the always-on voice IM mentioned in the previous section.

  Besides P2P communication in cars, telematics will likely become widespread, enabling the car to exchange messages with roadside or satellite systems for a wide range of applications from security and travel assistance to maintenance of the vehicle.

- **Laptop PCs** Increasingly supplied with built-in wireless communication capabilities, using technologies like Bluetooth and 802.11 (WiFi) will result in many of the traditional *local area network* (LAN)-based enterprise applications migrating to WAN. It is also likely that services will begin to truly combine the use of a cellular phone with a laptop computer. For example, simple wireless messaging applications such as SMS could be used for roaming users to request a login password when they want to start operating their WiFi-enabled PC from a new hotspot in a café, airport lounge, or hotel.
- **Additional nomadic devices** MP3 players, digital cameras, or PDAs will soon work in combination with cell phones via Bluetooth. This will enable users to use their cell phone as a wireless modem to handle file swapping with remote entities. A number of MMS pictures and music download services will make use of such combinations of separate devices rather than counting on the deployment of integrated combination products.

  Some consumer devices such as digital cameras or universal remote controls may be built with embedded wireless modems to enable those devices to communicate with online services. For example, interactive TV shows using SMS and MMS messaging to interact with viewers could exploit both standard phones as well as dedicated remote controls.

Availability and Speed of Networks

Many existing and key applications such as photo swapping, text or voice IM, and wireless e-mail will truly thrive with the advent of technologies such as GPRS and 1xRTT.

The history of new media, however, shows that the simple transposition of existing applications onto a new support is not usually the driving force. In almost all cases, it is creativity that triggers a brand new breed of services for the new medium. This makes it virtually impossible to predict which new applications will be created due to the arrival of faster, always-on networks; the only thing that can be predicted is that they will be numerous.

New Cultures and Geographies

Wireless messaging applications tend to be local and specific to a culture and environment. Unlike with the wired Internet, most users expect to see services on their cell phones in their own language and related to their immediate environment rather than some universal content.

Entirely new geographic areas and cultures will enter the wireless applications market in the coming few years, including China, Latin America, Central and Eastern Europe, India, and Africa. Applications for these markets and cultures have yet to be devised. It is expected that completely new ways of using wireless messaging will emerge to support local services in a way that meets the habits, cultures, and preferences of these regions and their people.

Greater ROI and efficiency gains can be achieved by extending traditional wireline applications wirelessly. We, as industry experts, would advice businesses to investigate how wirelessly enabling their applications could increase productivity and overall cash flow.

CHAPTER 5

Wireless Messaging Infrastructure

This chapter takes a closer look at the lower mechanics of wireless messaging and discusses the infrastructure required for messaging on a conceptual level. The objective of this chapter is to provide an overview of how a messaging network works. The messaging infrastructure required for *second-generation* (2G) messaging, or *Short Message Service* (SMS), will be discussed first, followed by a discussion on the *Wireless Application Protocol* (WAP). Figure 5-1 contains a high-level overview of the network elements involved in messaging over a wireless network.

Infrastructure Overview

Global System for Mobile (GSM) communications terminology has been used to describe the network elements. The high-level overview applies to all kinds of cellular networks, although the terminology used in *Code Division Multiple Access* (CDMA) and *Time Division Multiple Access* (TDMA) differs in some cases from the terminology used in this chapter.

Components of Wireless Messaging Solutions

Base Transceiver Station (BTS)

In order to provide wireless access to a wireless network, radio access is an obvious prerequisite. Radio access is provided by a network element called a *Base Transceiver Station* (BTS). A BTS is actually an antenna that is about 30 feet tall. A BTS provides radio access for wireless devices that are in the area of that particular BTS. Typically, a BTS covers an area (also referred to as a *cell*—hence the term *cellular network*) of several tens of kilometers, depending on the location of the BTS. In order to provide continuous coverage in a populated area, multiple adjacent BTSs are required.

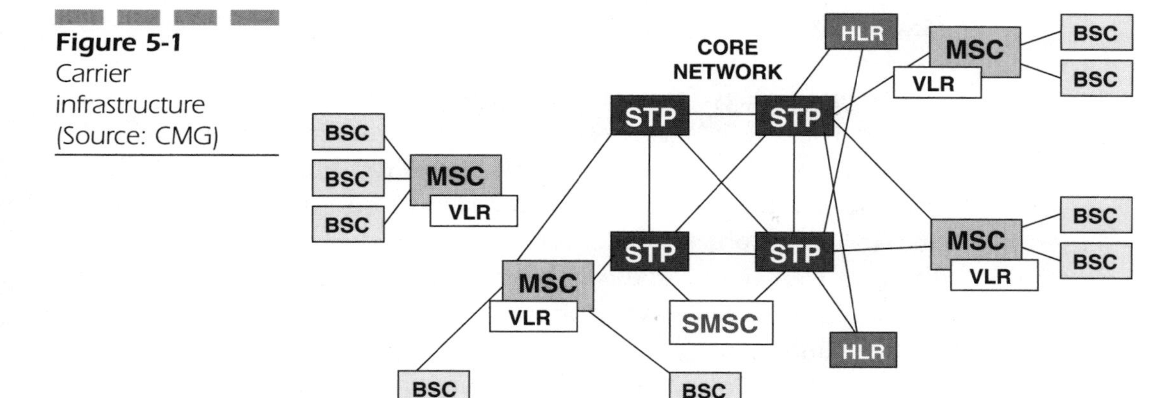

Figure 5-1 Carrier infrastructure (Source: CMG)

Maximum efficiency in coverage can be obtained if the BTSs are built in strategic locations, such as on top of high buildings in cities and outside the cities alongside highways. Nevertheless, in order to provide national coverage, a wireless carrier needs to deploy hundreds or even thousands of BTSs.

The degree with which a wireless carrier provides coverage used to be an important differentiator between competing carriers—hence the vast interest carriers have had in building BTSs in strategic locations in order to provide maximum coverage at minimum cost. Most countries in the Western world have at least three competing carriers, who are all in search for the most value for their money in terms of coverage. As a result, the management of many church societies have found a new source of income by selling building rights for BTSs on top of the tower of their churches.

Base Station Controller (BSC)

To use an analogy from the media, music radio stations that broadcast in the same geographical area use different frequencies for broadcasting their programs. This prevents them from interfering with other radio stations. However, music radio stations that broadcast in geographically different parts of the same country or in different countries can easily use the same *radio frequency* (RF) because no interference will appear between the broadcasts.

The same principle applies to the RFs used by adjacent BTSs—they are not allowed to overlap. Nonadjacent cells are allowed to reuse RFs, which explains the main advantage of a cellular network when compared to older analog, noncellular networks. Cellular networks cannot run full in terms of the number of subscribers, unlike analog networks, which can run full in terms of the number of subscribers (as each subscriber is assigned its own frequency).

Defining which BTS uses a particular frequency is called *radio planning* and is one of the most labor-intensive tasks of a wireless carrier. The RFs that are used in a particular geographically related group of BTSs are stored in a *Base Station Controller* (BSC).

As adjacent cells use different frequencies, when a cell phone user is traveling from one cell into the next cell, the cell phone needs to switch over automatically to a different frequency. This process is called *hand over*, which is one of the main tasks of the BSC.

Mobile Switching Center (MSC)

The *Mobile Switching Center* (MSC) is the central network element involved in all network traffic. Usually, MSCs are organized according to a larger geographical area. Each MSC connects to multiple BSCs that fall under the geographical hierarchy of the MSC.

For voice calls, the MSC is responsible for call setup and breakdown and potentially routing the call through to the MSC that is servicing the recipient phone. For messaging, the MSC's task is limited to relaying the submitted SMS through the core network to the SMSC, and vice versa.

In order to prevent unauthorized usage of the wireless network, other important tasks of an MSC include the registration and authentication of a cell phone. As the MSC is involved in all activities of a cell phone (for example, registering the cell phone on the network, making voice calls, and sending SMS), it also updates the *Visitor Location Register* (VLR) with the most recent location of the cell phone. For example, if a cell phone user drives 40 miles by car, he or she will be serviced by at least two MSCs, which take over the servicing of his or her cell phone during the ride. The MSC is notified by the MSC that was previously servicing that cell phone user when it should take over the servicing of a particular moving cell phone user.

Visitor Location Register (VLR)

As soon as an MSC finds out that a particular subscriber is in the geographical area it services, it will retrieve the subscriber data from the appropriate *Home Location Register* (HLR). The subscriber data is then copied to the VLR (the MSC's local location register). The VLR is usually integrated with the MSC and contains a copy of the data of the visiting subscribers who are serviced by the MSC at that point in time. At this point, the VLR can be updated with the most recent information that has been received by the MSC.

The VLR updates the HLR with the most up-to-date information. As a result, the VLR will always be more up-to-date than the HLR. This is why some carriers prefer to use the VLR's *presence* information on a particular subscriber instead of the HLR's information. The HLR is described in more detail in the section "Home Location Register (HLR)."

Roaming

One other major characteristic of a wireless network is that a cell phone can be reached at any location where there is coverage, either from the wireless carrier itself or from one of its partners. For instance, a Dutch GSM subscriber can travel with his or her cell phone to Malaysia and still receive voice calls and SMS messages. This is called *roaming*, which is covered by *roaming agreements* between carriers.

Roaming is an important source of income for a carrier. Whenever cell phone users set up or receive a voice call while they are not in the home network (that is, in the network of the operator that they have a subscription with), they are charged with roaming fees. Roaming fees depend on the agreement between the two operators. Recently, global carrier groups such as Vodafone and Orange have started to leverage the fact that they own a global network. In this case, roaming fees between two networks that belong to the same global carrier group are either very cheap or are not charged at all. For international corporations that use wireless cell phones for their employees, this is a reason to move all company subscriptions to a global carrier.

Home Location Register (HLR)

The obvious question is, how does the network know where the cell phone resides at any given moment in time? After all, the appropriate BTS needs to broadcast the voice call setup or SMS message to the right cell.

The answer to the mystery lies in the HLR. HLRs are organized according to number ranges and generally service up to 1 million subscribers. For a voice call, the MSC servicing the originator of the call will query the HLR of the recipient of the call to find out which MSC looks after the recipient at that point in time. In the example of the Dutch subscriber being telephoned in Malaysia, the Dutch HLR will be queried by the MSC servicing the originator of the call. This occurs regardless of where the call originated because the Dutch wireless number (+316xxxxxxxx) maps onto a specific Dutch HLR.

In order for a subscriber to be reachable on the wireless network, the subscriber's data needs to be stored in the HLR. The HLR stores a lot of information, but for our high-level overview, we will only

Table 5-1

HLR overview

Term	Description
MSISDN	Mobile Station ISDN—the dialed number, which is also referred to as the Directory Number.
IMSI	*International Mobile Station Identification*—the identification of the *Subscriber Identity Module* (SIM) card, a small chip card containing the subscription. Currently, this is only used in GSM and *General Packet Radio Service* (GPRS) networks.
Service level	Among others ■ Is the subscriber allowed to set up voice calls outside his or her own network? ■ Is the subscriber allowed to send an SMS? ■ Is the subscriber allowed to receive an SMS? ■ Is the subscriber allowed to send an SMS outside his or her own network? ■ Is the subscriber allowed to receive an SMS outside of his or her own network? ■ And so on
Servicing MSC	The MSC that services the subscriber at that particular time.
Cell ID	The identification of the cell where the subscriber resides at that particular time.

Source: CMG

describe the data stored on a subscriber relevant to back up the conceptual explanation.[1] See Table 5-1.

In GSM, three numbers are associated with a cell phone user: the MSISDN, the IMSI, and the *International Mobile Equipment Identity* (IMEI). The MSISDN is the dialed number. The IMSI identifies the SIM card, a small chip card containing the subscription. A SIM card can be removed from the cell phone and put in another cell

[1]The example is about GSM. In TDMA and CDMA, similar information is stored, but different terminology is used.

phone, thus making the cell phone user reachable with his or her own number using a different cell phone. Finally, the IMEI identifies the cell phone itself.

The question pops up why three numbers are used instead of just one? This is for two main reasons:

- **Wireless number portability** The MSISDN does not uniquely identify a subscription with a particular carrier in a wireless number portability environment. If a cell phone user wants to move to a different carrier while keeping his telephone number, he will keep his MSISDN, but he will receive a different SIM card and therefore also a different IMSI.
- **Antitheft protection** When a cell phone is reported stolen to the police, the IMEI and IMSI are blocked in the network on request of the police. As a result, neither the subscription nor the cell phone can be used by the thief, even if a different SIM card is used.

Signaling Transfer Points (STPs)

In order to provide fail safe, redundant, and still manageable connections between the various network elements described earlier, larger wireless carriers have organized their network using *signaling transfer points* (STPs). An STP's task is the telecom equivalent of routing traffic on the Internet. It also performs routing, address translation, and failover functionality. Instead of *Transmission Control Protocol/Internet Protocol* (TCP/IP), however, wireless network components use a protocol called SS7, which is also referred to as C7.[2]

In a network that is organized through STPs, the connections between them is crucial as they carry a huge amount of data since all traffic is transported through the STPs. This is why every STP is accessible by any other STP through multiple dedicated high-capacity access paths.

[2]Recent standardization enables SS7 over IP (SIGTRAN standard).

The recent success of SMS has put STP-centric infrastructures under extreme pressure because the enormous volumes of SMS messages are clogging up the networks. For this reason, carriers are currently investigating available alternatives.

A promising, cost-efficient alternative that provides much more bandwidth seems to be to migrate the inter-STP traffic to IP traffic using SS7 over IP SIGTRAN standards. Leading suppliers of IP-enabled STP infrastructure vendors are Cisco, Tekelec, HP/Compaq, and Airslide.

Short Message Service Center (SMSC)

The end user's perception of the *quality of service* (QoS) of the SMS service is determined by two factors:

- **The speed at which an SMS is delivered to the recipient** The delay introduced by the network between submission and delivery should be approximately seven seconds, provided the recipient is reachable at the time the message was submitted.
- **The availability of the SMS service** The SMS service should always be available.

The end-user perception of the quality of the SMS service is determined by the *SMS Center* (SMSC).

When a subscriber sends an SMS, the intended recipient of that SMS may not be available at that time, for example, because the recipient's cell phone has been switched off or is out of reach. The difference between sending an SMS and making a voice call is that the SMS will be delivered to the recipient as soon as the recipient subscriber switches on his or her phone again.

Again, this looks like a technical miracle. How does the network remember who sent a message and to whom it must be delivered while keeping track of when the intended recipient becomes available again?

Obviously, this must mean that the network temporarily stores the SMS somewhere, but the network should be intelligent enough to notice that the intended recipient has become available again. This principle is called *store and forward*. Whenever the recipient is

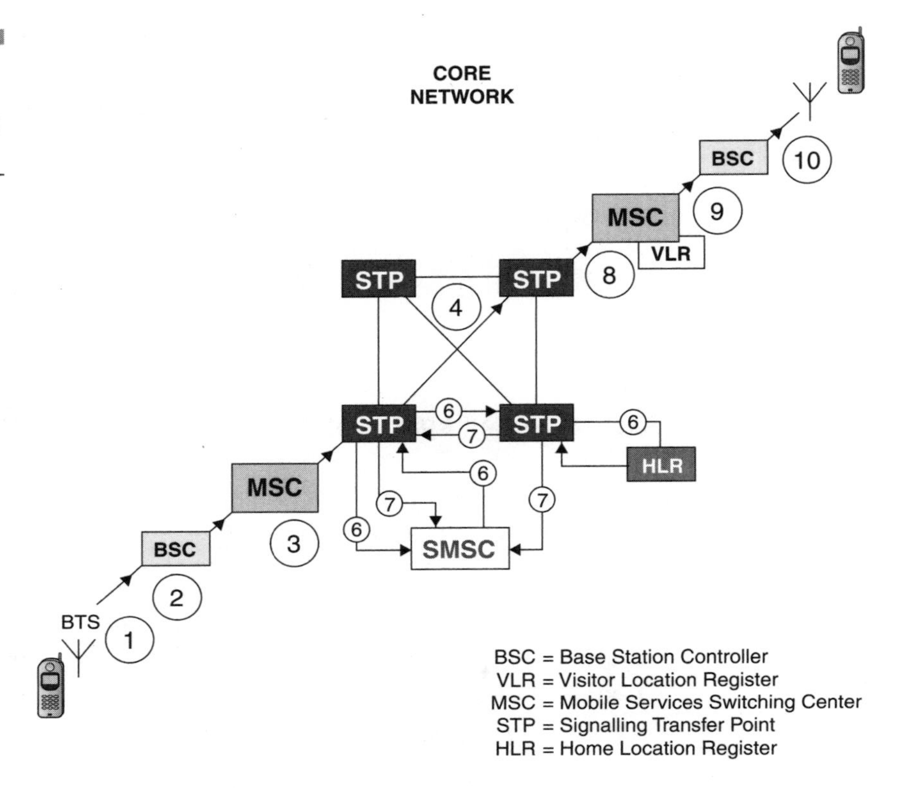

Figure 5-2 Further carrier infrastructure detail (Source: CMG)

under coverage of a roaming partner's network, the network should deliver the SMS to the recipient.

To make the previously described mechanism possible, an SMSC is required. The SMSC within a carrier's domain that will be used by a particular subscriber can normally be configured in the cell phone through the menu option *Service Center Address* setting.[3]

Figure 5-2 provides further detail into the core messaging network.

When a subscriber sends an SMS, the SMS travels through the BTS, BSC, MSC, and the core network, and is then stored in the SMSC. After the SMS has been stored in the SMSC, the message is

[3]The Service Center Address is configurable in GSM cell phones. CDMA and TDMA cell phones generally do not offer a similar self-configurable option.

accepted in the reverse order toward the sender of the SMS. When the SMSC has stored the SMS, the sender of the SMS will see *Message Sent* on the display of his or her cell phone. Novice SMS users generally tend to mistake the Message Sent display for the message being delivered at the recipient. It's important to realize that this is not the case.

After the SMS has been stored, the SMSC will immediately make an attempt to deliver the message to the intended recipient. To be able to deliver the message, the SMSC must first find out where the recipient resides at that point in time. For this reason, the SMSC will query the HLR associated with the recipient's number range. The HLR will respond as to whether the recipient's cell phone is switched on and is under network coverage. If the cell phone is switched on and under coverage, the HLR will indicate in its response to the SMSC that the MSC will service the recipient at that point in time. When the SMSC receives the response of the HLR, it will deliver the SMS toward the MSC indicated in the HLR's response. The MSC will check its local VLR for the most up-to-date location information, and subsequently the MSC will send the message to the indicated BSC, which will then forward the SMS to the BTS servicing the area where the recipient resides. The BTS will then transmit the SMS to the cell phone. Upon receipt of the message by the recipient's cell phone, the recipient's cell phone will confirm to the network that it has received the SMS. The response is relayed back through the network to the SMSC, which will then consider the SMS to be delivered and consequently delete the SMS from the SMSC.

If the recipient's cell phone is switched off, the HLR will inform the SMSC that the subscriber is not available. In this case, the SMSC will not continue delivery attempts for that particular recipient. Instead, the SMSC will request that the HLR be notified as soon as the recipient switches on his or her cell phone. If the SMSC is not notified by the HLR within two to three days (depending on the carrier's SMS service level), the SMSC will delete the message.

As soon as the recipient's cell phone is switched on, the MSC updates the VLR, the VLR will update the HLR, and the HLR will notify the SMSC that the recipient is available again. The SMSC will then immediately perform a delivery attempt as described previously.

How Trustworthy Is the SMS Service?

A subscriber who sends a message is in many cases interested to find out when the message has actually been delivered to the recipient's cell phone. To find out whether an SMS has actually been delivered to the end recipient, the sender of the SMS has the option to request a delivery confirmation by selecting a certain menu option on his or her cell phone. This mechanism is called *status report* and is a standard feature available in modern GSM networks. A status report will inform the sender whether the SMS was delivered immediately or whether the recipient's phone was unavailable at the time the message was submitted. In the latter case, a second status report will be sent as soon as the SMS has been delivered, which will occur as soon as the recipient phone switches on his or her phone. The status report is actually an SMS generated by the SMSC itself when it has received the result of the delivery attempt from the HLR or MSC and will be delivered to the cell phone of the sender of the original SMS.

Although the status report confirms delivery of the SMS to the recipient's phone, GSM currently has no mechanism to report to the sender that the message has actually been read.[4]

The Wireless Services Stack

The previous section described the mechanics of how SMS works. Now it's time to take a look at how SMS can be used as a bearer for *value-added services* (VASs), thus creating wireless services.

SMS is not only suitable for *person-to-person* (P2P) traffic, but it is also suitable for wireless access of VASs. Figure 5-3 gives an abstract overview of how VASs can make use of SMS as a bearer.

[4]In other cellular network types such as CDMA-One and CDMA2000, a *Message Read* report can be requested by the sender, which will report to the sender of the SMS when the recipient has actually read the message. The same principle has been made available in the *Multimedia Messaging Service* (MMS).

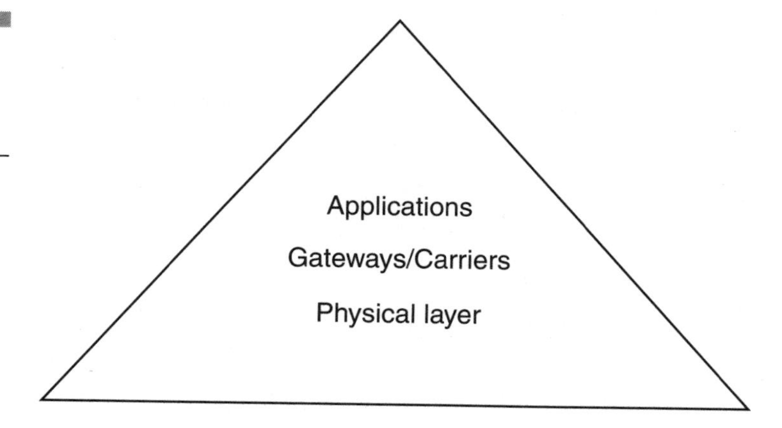

Figure 5-3 Value-added services (Source: EMC)

Applications

In this context, applications are services that can be accessed by end users and are tailored for the wireless market. Applications can be divided into the following types:

- **Wireless request/wireless response application types** With this application type, the VAS receives an SMS-based request for a service sent by a cell phone user. Upon receipt of the request, the application will create the content and send the content back in an SMS-based response to the requestor of the service. Typically, these kinds of applications are information services, such as airline information services (for example, if details of a certain flight number are requested, they are returned back in an SMS message) or stock quote information services. The end user is usually charged for the request for information service.
- **Web request/wireless response application types** With these application types, the end user does not know yet what kind of content he or she is looking for, so he or she needs a way of browsing through the available content. The browsing can usually be done through Internet web access, and when the content is selected, the content is then delivered to the cell phone

of the end user. These types of applications have become quite popular lately, especially applications that provide ring tones and picture messages (such as Smart Messaging or *Enhanced Message Services* [EMS]).

Because the content provider cannot charge for the (web-based) request, these kinds of applications require charging the recipient upon receipt of the content. The amount charged is dependent on the content that has been purchased. This situation can be compared to a real-life shop, where you do not have to pay for entering the shop, but must pay for the goods that you have purchased in the shop.

Charging the recipient upon receipt of the content is called *reversed charging*. Reversed charging can be considered a prerequisite for applications of this type and will become much more of a requirement for wireless carriers in order to tie in these kinds of application providers.

- **Wireless input/mass-medium output application types** With these application types, many end users send content toward the application, and the content is displayed in an aggregated or nonaggregated form on some mass-medium output channel. A popular example of this type of application is interactive TV. The concept of interactive TV has been around before by attempting to have the public participate in a program through the Internet. However, web-based interactive TV turned out to be impractical. After all, how many people will watch TV with a portable computer on their lap? In contrast, the use of SMS for public participation in interactive TV shows has turned out to be a huge success. TV viewers find it much more convenient to use their wireless cell phones to participate in a TV show rather than using a computer connected to the Internet.

  The use of SMS as a medium for interactive TV really puts the network to the test since large volumes of SMS will be sent in a very short timeframe. The worst thing that a TV production company can do is to start an interactive TV show using SMS without joint planning with the wireless carrier. These kinds of SMS peaks may result in overload situations if the carrier does not have sufficient SMSC capacity, resulting in a bad user experience for the participating TV viewer. You only get one chance to make the right impression so QoS is everything in the carrier world.

Past experiences with interactive TV in Europe have proved that joint planning among the TV production company, the TV station, and the carrier is the most effective way to provide a superior user experience. Interactive TV using SMS is normal day-to-day business in Europe. Today's examples of interactive TV using SMS include *Big Brother*, *Who Wants to Be a Millionaire*, *MTV Clip Requests*, and many more.

- ***Mobile commerce* (m-commerce) application types** With m-commerce, the phone subscription becomes a credit card or a debit card (in the case of prepaid cell phones) in which services or goods can be paid for.

  When a consumer chooses to pay for goods in a shop using m-commerce, the shopkeeper will trigger the m-commerce application to send an SMS to the cell phone of the end user. The m-commerce SMS will state the price that needs to be paid. When the user replies to this SMS with an electronic signature (that can be stored in the cell phone by a SIM toolkit application), the cost of the goods will be added to the phone bill. In the case of a prepaid subscription, the cost can be deducted from the prepaid balance. Similar to a credit card company, the carrier calculates a percentage for providing the m-commerce service.

  To prevent queues at the shop counter, high interactivity between the m-commerce application and the consumer's cell phone is absolutely essential. For the carrier, this means that the SMSC must allow for high interactivity and that the SMSC must be highly available. An SMSC that is temporarily out of service will hamper m-commerce transactions, causing loss of revenue for both the shopkeepers and the carrier itself!

Gateways/Carriers

Regardless of the application type, a company that runs a business based on wireless applications will require access to all wireless subscribers in one or multiple countries, regardless of the carrier that the wireless subscribers have a subscription with. However, from a wireless application company's point of view, making agreements with individual carriers is time consuming and seriously hampers the time to market of a new service.

To overcome this hurdle, a number of companies are providing gateway services. The gateway services already have connections to each carrier in one or multiple countries, which simplifies the life of an application provider company as it only needs to make commercial agreements with one company: the gateway provider. This is illustrated in Figure 5-4.

Virtual Mobile Applications

Recently, an alternative for nationwide application access has arisen—the Virtual Mobile. The Virtual Mobile is a network element simulating a cell phone. Because cell phones are reachable from any GSM network in the world, a Virtual Mobile is capable of offering worldwide access to a particular application. Virtual Mobile solutions are

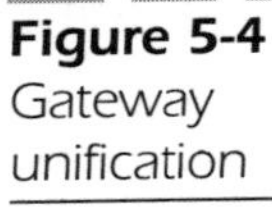

Figure 5-4
Gateway unification

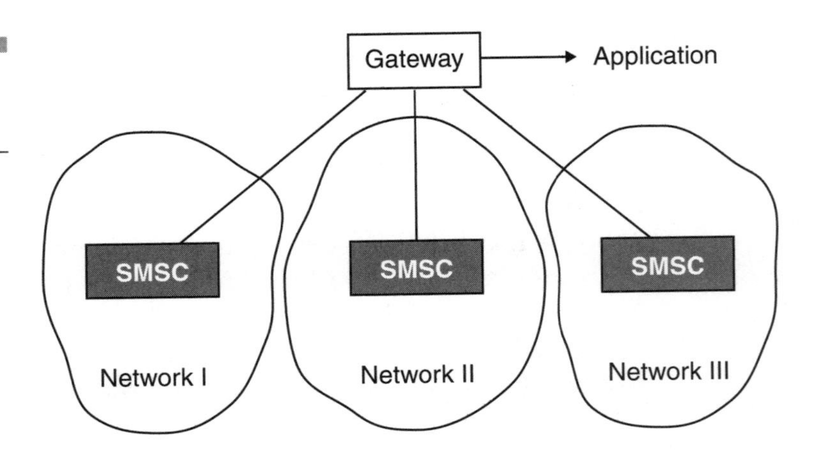

offered by a few specialized vendors and also by most of the main SMSC vendors.

Although the Virtual Mobile can be a good solution for low-traffic applications, for high-traffic applications, the Virtual Mobile concept introduces a dangerous situation where networks can get overloaded, causing the entire SMS service to suffer.

As stated in the previous section, most application providers require at least nationwide coverage of their services. First, this requires a service that is accessible through SMS since 100 percent of the GSM subscriber base is SMS enabled. Second, it requires network access from all GSM networks in the country and even outside of the country.

Carriers have already offered a similar, familiar technique. After all, it is already possible to send SMS from any GSM network to another mobile in any other GSM network; therefore, the solution for worldwide SMS service was conceived to be an application simulating a (virtual) mobile in the network.

Meanwhile, on the network side, another reason for developing the Virtual Mobile has resulted in SMS interconnect contracts between carriers. Historically, carriers with SMS interconnect agreements (that is, the possibility to send SMS between the networks) made no special arrangements for relaying SMS into other networks. Carriers considered the number of SMS messages coming into their networks to be more or less equal to the amount of SMS message being sent from their own network. However, with the introduction of worldwide accessibility of services, such as free SMS messages being sent from a carrier's web site, the balance was disturbed. A good example is the MTN service provider, SMS web service, which is heavily used by end users around the world. These SMS messages can be sent from the South African MTN network to all GSM interconnect partners. It goes without saying that the amount of *mobile-terminated* (MT) messages received from the MTN network far exceeds the number of messages being sent toward the MTN network. This especially accounts for the networks with the largest subscriber bases.

To resolve this issue, carriers are currently planning to establish charged SMS interconnect agreements where the sending network carrier has to pay relay charges. This centers on the fact that on a monthly basis the number of messages received from a certain network is balanced with the messages being sent toward that particular network. The net difference is then rated against a previously agreed amount per message, such as 5 Eurocents per message. These SMS interconnect agreements have recently been established in U.K. networks. Due to this arrangement, carriers with a charged SMS interconnect agreement have a vast interest to attract more MT messages into their network than have MT messages going out of their network.

At first sight, the conclusion would be that a Virtual Mobile application has a double business case for the hosting carrier. Additionally, it also has a direct business case for the SMS interconnect partner (an SMS sent from the interconnect partner's network can be charged directly to the subscriber). This sounds really too good to be true. Is it a win-win-win situation? Let's take a closer look at the mechanics of the Virtual Mobile.

Limitations of the SMSCs A throughput bottleneck for Virtual Mobile applications is generally located at the SMSC of the SMS interconnect partner(s). To understand this, a brief description of the internals of an SMSC is necessary.

An SMSC is used to send and receive SMS messages for a number of subscribers or applications. Because storage capacity is limited, the number of SMS messages that can be stored, which is the product of the number of subscribers and the message queue size per subscriber, is limited as well.

When this queue (also known as the *recipient buffer*) is full, other messages for the same recipient will be rejected.[5] Typical operational capacity for regular numbers varies between 25 and 50 buffered message capacity per mobile recipient.

[5]The recipient buffer applies to CMG's SMSC only. SMSCs from other vendors use databases that are addressed using memory-cached *linked lists*. If these linked lists exceed certain limits (an approximate magnitude of 10,000 messages), the entire SMSC could malfunction or crash.

Because the sending SMSCs do not recognize that the Virtual Mobile is actually a high-capacity recipient, the recipient queue for the Virtual Mobile at the sending SMSC will become full because of limitations in the network.

Limitations of the Network From an SMSC's point of view, sending SMS messages to a mobile involves stopping and waiting. The receiving mobile must confirm receipt of an SMS before a second (and third and so on) can be sent. If a service is characterized by bursty message traffic, such as an SMS voting application, this would lead to a massive rejection of messages, or, in the worst case, a system crash of SMSCs.[6]

Trying to solve this issue by enlarging the size of the recipient buffer is not under the hosting network carrier's control and is not a solution to the problem. Even under the assumption that this buffer could contain thousands of messages, the messages would be delivered to the application with an unacceptable delay.

This delay is caused by transition times of signaling traffic passing through STPs and (national or international) gateways, and the time needed to interrogate the HLR and processing time within the Virtual Mobile application for generating the acknowledgement. This delay is even enlarged because the acknowledgement must follow the same path (excluding HLR interrogation).

Typical network delay times for SMS are between 6 and 10 seconds (based on national traffic—international traffic most likely has higher figures). This comes down to a maximum number of messages sent by an SMSC to the Virtual Mobile of 600 messages per hour!

Virtual Mobile SMS Voting: A Messaging Nightmare? Given the interconnect business case, a carrier is interested in attracting as many MT messages as possible without sending back a reply to

---

[6]CMG's SMSC has been designed not to crash under such circumstances. It keeps rejecting messages until the buffer is under its maximum size. As soon as the number of messages in the buffer drops below the maximum, messages are accepted again.

the sender of the MT message. The first application that springs to mind that will attract a lot of MT traffic sent to the Virtual Mobile is an SMS-based voting application where the voting number is broadcast on a TV show.

Under the limitations described previously, it requires little imagination to understand what would happen if a carrier hosted an SMS voting application where the number would be broadcast during a prime-time TV show. Experience with SMS voting in Germany has shown that additional peaks above 200 messages per second per participating GSM network are to be expected. With a network delay of six seconds per message, the sending SMSCs will block within a fraction of a second. The public will think that the service does not work, and TV viewers will start calling the carrier's customer-care help desk.

The hosting carrier will probably be sued for liability, and the SMS interconnect agreements will be seriously endangered, thus seriously endangering the overall success of SMS in a country. For this reason, the connectivity provided by a gateway toward applications should be preferred over Virtual-Mobile-based application connectivity.

Physical Layer: Accessing the Network

As described in the section "Components of Wireless Messaging Solutions," submitting an SMS and delivering it to the recipient is handled through an SMSC. This also applies for application traffic. However, an application can bypass the complete core network and access a carrier's SMSC directly through a TCP/IP link (provided that the proper arrangements have been made). As described in the section "Gateways/Carriers," in most occasions, the TCP/IP link toward the SMSC is connected to a gateway, which in turn also connects to SMSCs of other carriers. Generally, the TCP/IP connections toward a carrier's SMSC are in a private IP domain, thus protecting the SMSCs from hackers.

Wireless Request to Application

A wireless request to an application differs from normal P2P messaging. When a subscriber sends a wireless request to a particular application, the message is generally sent toward an easy-to-remember short code address. As with P2P messaging, the SMS travels through the BTS, BSC, MSC, and the core network, and is then stored in the SMSC. After accepting the SMS toward the cell phone user (the cell phone displays *Message Sent*), the SMSC will translate the short code entered by the sender to the application's TCP/IP connection and deliver the SMS-based service request over the TCP/IP link through the gateway toward the application.

Wireless Application Response

When an application sends an SMS-based response, the SMS is relayed through the SMS gateway and is stored in the SMSC. After the SMSC has accepted the SMS toward the application, the SMSC will immediately make an attempt to deliver the message to the recipient's cell phone. The same mechanism applies to the delivery of a wireless application response as that of P2P messaging, as described in Figure 5-1.

An application generally requests for delivery confirmation to make sure the content has been delivered to the subscriber. The SMSC will inform the application whether the SMS was delivered immediately or whether the recipient phone was unavailable at the time the message was submitted. In the latter case, the application will be notified as soon as the SMS has been delivered, which will occur as soon as the recipient switches on his or her phone.

The Critical Role of the SMSC

As described in the previous section, an application can directly access an SMSC over a private TCP/IP connection. In some European countries, because of historic reasons, older data networks, such as X.25 networks, are used to connect an application to an SMSC.

An application developer obviously needs more details than just having to use TCP/IP connections; how an SMS needs to be formatted over the TCP/IP connection must also be defined. The formatting of the SMS data itself is referred to as the *SMS application protocol*.

Originally, each SMSC vendor defined its own SMS application protocol and tried to promote it to application developers. SMSC vendors understood early that it would be in their best interest to try to attract as many application developers as possible, as this would increase the attractiveness of their SMSC product to the carrier. For this reason, the SMSC vendors made their version of the SMS application protocol publicly available and supported application developers with programming *application programming interfaces* (APIs), which can be downloaded for free off their web sites.

As a result, a number of competing SMS application layer protocols are in use in the world. Today, most SMSC vendors support one or more SMS application protocols in order to be able to connect as many SMS applications as possible to their installed SMSC sites.

Connecting to the Carrier

An application developer is forced to use the SMSC vendor-specific application protocols before the application can be launched to the wireless world, which is determined by the brand of SMSC. Depending on the SMSC that the application needs to connect to, the application developer has to choose one or the other application protocol. Some application developers even have to use two or more SMSC vendor-specific protocols.

Short Message Peer to Peer (SMPP)

Short Message Peer to Peer (SMPP) was originally defined by the Irish-based company Aldiscon, which was later acquired by Logica. The use of SMPP became widespread in regions throughout the world, with the exception of western Europe. Ownership of the

SMPP protocol was taken over by the SMPP Forum in 1997, which later renamed itself the SMS Forum.

The most common SMPP version in use today is version 3.4, which is supported by most major SMSC vendors. The SMPP 3.4 specification as well as a software developer's kit can be downloaded from the SMS Forum's web site at www.smsforum.net.

Universal Computer Protocol (UCP)

The *Universal Computer Protocol* (UCP) was originally defined by the *European Telecommunications Standards Institute* (ETSI) standardization committee for ERMES paging networks, UCP is the leading application protocol used in western Europe. Since western Europe boasts the world's most successful SMS market, UCP must be considered one of the two most important application protocols for carriers to adopt next to SMPP. The most common version of UCP in use today is version 3.5, which is supported by most major SMSC vendors. CMG released UCP 4.0 in early 2001, which allows content-based reversed charging. Because content-based reversed charging is a prerequisite for web request/wireless response application types, the popularity of UCP 4.0 is expected to increase rapidly in the coming years.

Open Interface Specification (OIS)

The *Open Interface Specification* (OIS) was defined by the Anglo-French-based company Sema Group, which was acquired by Schlumberger in early 2001. The use of OIS is not as widespread as SMPP or UCP, but its significance must not be underestimated. In some cases, OIS is the only SMS application protocol available to access Vodafone networks.[7]

The most common version of OIS in use today is version 5.1, which is only supported by SchlumbergerSema's SMSC and CMG's SMSC.

---

[7] Vodafone currently owns the world's largest network.

Computer Interface to Message Distribution (CIMD2)

Computer Interface to Message Distribution (CIMD2) was defined by the well-known Finnish-based company Nokia. CIMD2 is used in few networks. Since Nokia's SMSC also supports SMPP and UCP, it is considered less important to implement CIMD2 for application developers. CIMD2 is supported by Nokia's SMSC and CMG's SMSC. The specification of CIMD2 can be downloaded from Forum Nokia's web site at www.forum.nokia.com.

Telocator Alphanumeric Protocol (TAP)

The *Telocator Alphanumeric Protocol* (TAP) was originally defined by PCIA for American two-way paging networks. TAP needed to be supported by SMSC vendors for two reasons:

- To create the possibility of interconnect between SMS and paging networks in the United States
- To support SMS access for existing applications that were originally written for paging networks.

Although TAP is supported by most major SMSC vendors, it should be considered to be a legacy protocol since it does not support specific SMS features that are needed to send ring tones, pictures, and so on. For application developers aiming strictly at the SMS market, the TAP protocol is not considered a logical choice to implement.

Simple Mail Transfer Protocol (SMTP)

The *Simple Mail Transfer Protocol* (SMTP) was defined by the *Internet Engineering Task Force* (IETF) for the purpose of e-mail. Native SMTP access on the SMSC is generally not supported by carriers for a number of reasons:

- **Charging** It is not possible to charge the sender of an SMS if the message is submitted through the Internet.

- **Security** Connecting an SMSC to the Internet poses the risk of a denial-of-service attack on the SMSC. As described in previous sections, an SMSC is a business-critical device and unavailability of an SMSC means loss of revenue. Connecting an SMSC to the Internet is therefore not allowed by carriers.
- **Functionality** SMTP was not designed for sending an SMS. Like TAP, SMTP does not support the SMS features necessary to send pictures and ring tones.

For application developers who want to make a serious business from SMS-based services, SMTP should not be considered a suitable option.

QoS of SMS

The perception that an end user has of the quality of a VAS application not only depends on the quality and speed of the application itself, but it also depends on the underlying wireless carrier's network. For this reason, application providers are particularly interested in the service level of the underlying network, which is referred to as the QoS level. The QoS level is usually part of the *Service Level Agreement* (SLA) between the carrier and application provider. The carrier provides a high QoS level to the application provider if the carrier's network does not decrease the service level of the application.

Direct QoS Level

Direct QoS is determined by the following factors:

- The delay introduced by the carrier's network between a wireless service request being submitted by the cell phone user and the receipt of that service request at the application.
- The delay introduced by the carrier's network between a wireless service response being submitted by the application and the receipt of that response by the cell phone user.

- The capacity of the wireless network, for example,
  - How many wireless service requests sent by cell phone users can be relayed to the application provider per second?
  - How many wireless responses can be sent by the application provider to cell phone users per second?
- The degree with which an application has control over a submitted message, for example,
  - Can the application receive a delivery confirmation or, alternatively, a nondelivery notification?
  - Can the application modify a submitted message?
  - Can the application delete a submitted message?
  - Can the application assign a certain maximum validity period to a message? For example, if the message has not been delivered before a certain time, the message will be considered outdated and will be deleted from the SMSC.
- Reversed real-time charging capabilities for prepaid users. In the case of submitting valuable content to prepaid users, is the prepaid user's credit checked and deducted with the proper amount before the valuable content (such as a ring tone) is delivered to the end user?

The following are key factors that influence the QoS of a network:

- The wireless coverage factor.
- The handsets in use in a carrier's network must support two-way SMS. The user-friendliness of the SMS interface on the cell phone is of extreme importance. Users with cell phones with a good SMS interface send more SMS messages than users with a less user-friendly interface. Nokia phones cannot be left unmentioned as an example of a cell phone with a good SMS user interface.
- The peak capacity of the core network; in other words, how many SMS can be relayed through the core network during the busiest time of the day with regard to SMS traffic?
- Performance of prepaid billing systems; if real-time credit validation is performed for every SMS, the transactional capacity

of the prepaid systems must allow for the same amount of peak SMS traffic as the core network and the SMSC must allow for (taking into account the amount of prepaid versus postpaid messages).

- The performance of the SMSC; in other words, how many SMS messages per second can be accepted and forwarded by the SMSC?
- The availability of the SMSC.
- The processing speed of the SMSC, that is, the delay introduced between receiving an SMS from the sender and forwarding it to the recipient.
- The bandwidth available on the TCP/IP link between the application and the SMSC.
  - How many SMS messages per second can be submitted by the application into the SMSC?
  - How many SMS messages per second can be delivered to the application by the SMSC?

Indirect QoS Reporting Facilities Required

A more subtle level is the indirect QoS. The QoS level required by an application provider is usually part of the commercial contract between the application provider and carrier. To prove that the required QoS level is actually being provided, the carrier needs to deliver a QoS report on the performance of the network to all the application providers. This is the indirect QoS level offered by the carrier to the application provider.

To provide such reports, a carrier needs strong reporting facilities that can measure the QoS of the carrier's network in almost real-time without degrading the performance of the SMSC. The challenge is that the amount of data is so huge (some carriers report more than 30 million messages per day), that business intelligence/data warehouse techniques are required to supply the QoS reports in time for all application providers. Although they fulfill an important business need for carriers, off-the-shelf reporting solutions for messaging have only recently come into the market.

Major SMSC Providers

The major (core) SMSC vendors in the market are Logica, CMG, Comverse, and SchlumbergerSema.

Logica

Logica is a leading global solutions company providing management and IT consultancy, systems integration, products, services, and support. Its Mobile Networks arm offers messaging and payment platforms. Logica's clients operate across diverse markets including the telecom, financial services, energy and utilities, industry, distribution, transport, and public sector markets. The company operates in 34 countries worldwide. Founded in 1969, Logica is the largest IT services company listed on the London Stock Exchange. More information can be found at www.logica.com.

Through the acquisition of Aldiscon in 1998, Logica became one of the main players in SMSC business. Because of its early worldwide presence, Logica inherited a worldwide installed base of SMSCs from Aldiscon. With the exception of Japan, where Logica is the only foreign SMSC vendor, Logica's SMSCs are generally installed at smaller networks with relatively low SMS traffic. Logica's SMSC, branded as Telepath SMSC, is feature rich. Due to Aldiscon's early global presence, Telepath SMSCs provide connectivity to most wireless network types, voicemail systems, and application protocol types.

Although they are quite suitable for low-traffic networks, Logica's Telepath SMSCs do not have the stability necessary to cater to continuous high-volume traffic scenarios. Additionally, when these SMSCs are running in *classic mode*, database maintenance is required every night during low-traffic hours. For operators, this is not ideal because when SMS traffic starts to pick up in a network, the low-traffic hour classification becomes less and less until finally there are no low traffic hours left for the maintenance required.

The original database-oriented design inherited from Aldiscon was later optimized by Logica with the introduction of *premium messaging mode*, which does not require database maintenance. However, while running in the enhanced mode, high traffic periods still pose stability issues.

CMG

CMG, which was established in 1964, is a full-service provider of management consultancy, systems development and integration, and outsourced management of targeted business processes. CMG has two divisions: *IT and telecommunications services* (ITC) and *Wireless Data Solutions* (WDS). WDS has gained global acceptance in wireless messaging, mobile Internet, and customer care and billing. It was built primarily on its delivery of high-performance, high-reliability SMSC to mobile network operators. Its portfolio today also includes WAP and i-mode connectivity systems, specialist mobile application servers, *Multimedia Message Centers* (MMSCs), *Unified Messaging* (UM), and systems for billing and customer management and development. With over 13,000 employees, CMG implements and supports applications for customers worldwide from bases in more than 20 countries. CMG is listed on both the London and Amsterdam Stock Exchanges. More information on CMG is available from www.cmg.com/wds.

CMG WDS has been involved in mobile messaging since the early 1990s when it was asked to design and develop an SMSC for a Nordic consortium. Unlike Aldiscon (later Logica), CMG originally did not have worldwide presence. It was active mainly in western Europe through its consultancy business.

The CMG SMSC is a multinode, fully redundant system. Due to the success of SMS in western Europe, CMG's SMSC is now in use in the largest GSM networks and became a business-critical device quite early in its product life cycle. Rather than focusing on features, CMG focused on stability and performance.

Comverse

Comverse, a unit of Comverse Technology, Inc., is a leading global provider of software and systems enabling network-based multimedia enhanced communications services. It offers solutions including call answering with one-touch call return, SMS, IP-based UM, MMS, instant communications, wireless information and entertainment services, voice-controlled dialing, messaging and browsing, prepaid wireless services, and additional *personal communications services* (PCS). Other Comverse Technology business units include Verint Systems, a leading provider of analytic solutions for communications interception, digital video security and surveillance, and enterprise business intelligence; and Ulticom, a leading provider of service-enabling network software for wireless, wireline, and Internet communications applications. Comverse Technology is an S&P 500 and Nasdaq-100 Index company. For additional information, visit Comverse's web site at www.comverse.com.

Comverse's main area of business was and still is selling voicemail systems. The first killer application for the use of SMS was to use SMS as a notification mechanism to inform the subscriber that voicemails were stored in his or her voicemail box. Originally used as a voicemail notification system, Comverse expanded its voicemail notification system to a full-blown SMSC, which it has branded as the I-SMSC. Comverse entered the SMSC market quite late, but its systems were rather low priced, which explains why Comverse has a significant market share.

Like Logica's Telepath SMSC, Comverse's I-SMSC is mostly used in lower-traffic networks. Since Comverse's I-SMSC is built around a database, it also requires database maintenance during low-traffic hours. Unfortunately, database-centric SMSCs are limited by the performance of the database engine; therefore, Comverse's I-SMSC is not suited to high-traffic networks.

SchlumbergerSema

Schlumberger is a global technology services company with corporate offices in New York, Paris, and The Hague. Schlumberger was founded in France in 1927 and has more than 80,000 employees in nearly 100 countries. The company consists of two business segments: Schlumberger Oilfield Services and Schlumberger Network Solutions and SchlumbergerSema. Schlumberger Network Solutions provides IT connectivity and security solutions to the E&P industry and a range of other markets. SchlumbergerSema provides IT consulting, systems integration, managed services, and related products to the oil and gas, telecommunications, energy and utilities, finance, transport, and public sector markets. More information on SchlumbergerSema can be found at www.slb.com.

Schlumberger acquired Sema in early 2001. Like CMG, Sema started its SMSC business in western Europe by building an SMSC for the Vodafone group, which it branded SMS2000. Like CMG's SMSC, SMS2000 is a fully redundant system that is used in high-traffic networks.

Wireless Gateways

What Are They?

A Wireless Gateway is a central SMS router, typically used for intercarrier SMS, that connects to each carrier's participating SMSC using wireless-technology-independent protocols such as SMPP. The individual carrier's SMSCs are configured to route messages that are sent to recipient addresses outside of the carrier's own network to the Wireless Gateway. The Wireless Gateway then determines, based on the recipient's number lookup into the assigned number ranges, which SMSC should deliver the message to the end recipient.[8]

[8]Number ranges in the United States and Canada are assigned in 10,000 number blocks called NPA-NXX where NPA corresponds to the *Numbering Plan Area* (or area code) and NXX corresponds to a particular switch. The NPX-NXX is the first part of the NPX-NXX-XXXX number as defined in the United States and Canada. The NPA-NXX numbering plans are compliant with *International Telecommunications Union-Telecommunications Standardization Sector* (ITU-T) E.164, *The International Public Telecommunication Numbering Plan*. In the United States, thousand-number pooling requires wireless carriers to implement a technology by November 24, 2002 that enables exchanges or numbers to be assigned to carriers in blocks of 1,000 instead of 10,000, thus yielding an efficient allotment and use of numbers by reducing the amount of stranded numbers.

Standardization of Wireless Gateways is coordinated by the SMS Forum, which owns and maintains the SMPP protocol. The *Cellular Telecommunications Internet Association* (CTIA) and *Canadian Wireless Telecommunications Association* (CWTA) also play a major coordination role in the standardization of Wireless Gateways in North America.

U.S. Market Importance

In contrast to Europe, within the United States, national SMS interconnect has not been easy since the wireless networks do not support the routing of SMS between the different technologies. Wireless Gateways are essential for the success of national intercarrier SMS in North and South America.

Although SMS has been widely available to North American wireless subscribers since 2000, it was not viewed as a usable medium for normal P2P communication until intercarrier SMS was established in mid-2002. After all, subscribers will not use SMS for P2P communication if it is not reasonably certain that the message can be delivered. The fact that the cell phone displays *Message Sent* even though the carrier's SMSC does not deliver the message without notifying the sender has been a past detriment for the SMS user experience in the United States. ("This stinks . . . the network told me that my message was sent and the recipient claims he never received my message").

Looking at experience in other parts of the world where intercarrier SMS interoperability was established (comparable figures in Europe showed an increase in chargeable message traffic as a result of SMS interoperability up to 900 percent), U.S. carriers have shown a need to resolve the SMS interconnect issue. The various industry groups such as the CTIA and CWTA have both endorsed the Wireless Gateway model as the most logical and workable solution.

The Interoperability Question

By mid-2001, it was clear that SMS had been an underperformer for the markets in the Americas. More importantly, however, even if an end user in the United States had a two-way SMS-enabled phone, he or she was not able to send SMS messages directly to other mobile users if the intended recipient did not subscribe to the same carrier. Other issues blocking interoperability consisted of

- Incompatible carrier network air interfaces (CDMA, TDMA, GSM, and *Integrated Digital Enhanced Network* [iDEN])
- Incompatible signaling standards among air interfaces (IS-41/IS-136 for TDMA, MAP for GSM, and IS-41/IS-95 for CDMA)

The handset of the end user would present the option of sending a direct SMS text message; however, due to the incompatibility of SMS infrastructures in North and South America, the user would enter the message, hit Send, and assume that it is safely on its way. The reality was that the message was being dropped because the sender's wireless carrier's SMSC did not relay messages sent to subscribers from a different carrier.

The problem was further compounded by the lack of full interoperability within some of the many large carriers in North America. These carriers are the products of multiple mergers and acquisitions, and the signaling interoperability among properties was not always complete. This further limited end users to messaging directly only within their own specific property.

Each of these barriers presented a confusing messaging experience and prevented carriers from establishing a consistent brand identity. As a result, end users in North and South America assumed that SMS was either too cumbersome or simply did not work.

To fully realize the promise of SMS messaging in the Americas, carriers had to convince their subscribers that SMS messaging is simple, reliable, and fast. One alternative, text messaging via e-mail (for example, through WAP), can be used to send a message to a mobile subscriber of a different network; however, there are several disadvantages for the end user, which explains why text messaging via e-mail does not provide the same ease of use as SMS:

- The originator must know the destination and sometimes confusing e-mail address, such as 1234567890@mobile.xyzwireless.net.

- The originator must know the recipient's wireless carrier.
- The message, which will appear on the recipient's handset, may contain an e-mail header, thereby limiting the amount of characters available for writing the true message (SMS messages are typically limited to 120 to 256 characters).
- The message may be truncated upon receipt, thereby destroying the integrity of the original message, depending on the configuration of the wireless carrier's SMSC.
- The originator may have a difficult time composing the SMS because many handsets with predictive text software limit its application only to direct SMS text messaging (for example, the predictive text software does not work for e-mail).
- The sender will not know if his or her message has been delivered. Although wireless carriers' SMSCs could be set up to provide Failed Delivery auto-replies (similar to Undeliverable messages received after sending an e-mail to the wrong address), most wireless carriers chose not to do so since such auto-replies are difficult to bill for and use signaling channel bandwidth.
- Transmission via e-mail protocol is slow. Even if a message is eventually delivered, the time between sending and receiving that message is entirely unpredictable and can range anywhere between 10 seconds to several hours.

Intercarrier Message Exchanges

Despite the strong business case for wireless carriers to establish SMS interconnect, the intense competition and lack of trust between wireless carriers has been a barrier to direct negotiation on SMS interconnection. To fulfill the market need for SMS interconnect, a number of companies have started a business that provides an SMS interconnect service called *intercarrier message exchanges*. In the United States and Canada, the CTIA and CWTA have promoted standards, guided by key carriers, for intercarrier message exchanges and gained technical support from the SMS Forum. In mid-2001, Canada was the first to develop a selection process to identify a single intercarrier SMS vendor to provide intercarrier SMS for all Canadian carriers. The CTIA began promoting intercarrier SMS in late 2001 with support from the SMS Forum.

A company providing an intercarrier message exchange service can use a Wireless Gateway to connect to the carrier's SMSCs through any IP-based connection such as T1s, Frame Relay, and even *virtual private networks* (VPNs). Intercarrier message exchanges generally charge the connecting carriers a fee per message. Consequently, they have made a strong business case out of generating more SMS business for the carriers.

In North America, the market is now evolving toward a full, national interconnection through a limited number of intercarrier message exchanges. The following are the major players as of June 2002:

- **MobileSpring/Verisign (partnership)** Provides the SMS interconnect service between SprintPCS, Cingular Wireless, and Numerex
- **Inphomatch** Provides the SMS interconnect service between AT&T Wireless, Verizon Wireless, Leap Wireless, Voicestream Wireless, and many smaller carriers
- **Wireless Services Corporation** Provides the SMS interconnect service for Nextel
- **CMG** Provides the SMS interconnect service for four Canadian operators: Telus Mobility, Rogers AT&T, Microcell, and Bell Mobility

The model for South America is slanted toward the carriers themselves providing intercarrier SMS by deploying a Message Gateway and connecting to one or more other carriers. Slowly, this network of gateways will be connected, forming another continental SMS network. Figure 5-5 illustrates the major messaging players and how they collectively make wireless messaging possible.

Since mid-2001, the intercarrier message exchanges have been in the process of interconnecting with each other so that full SMS interconnectivity will be established between the subscribers of the participating carriers. As can be seen in Table 5-2, the interconnection agreements already allow for an interconnect percentage of 83 percent in North America.

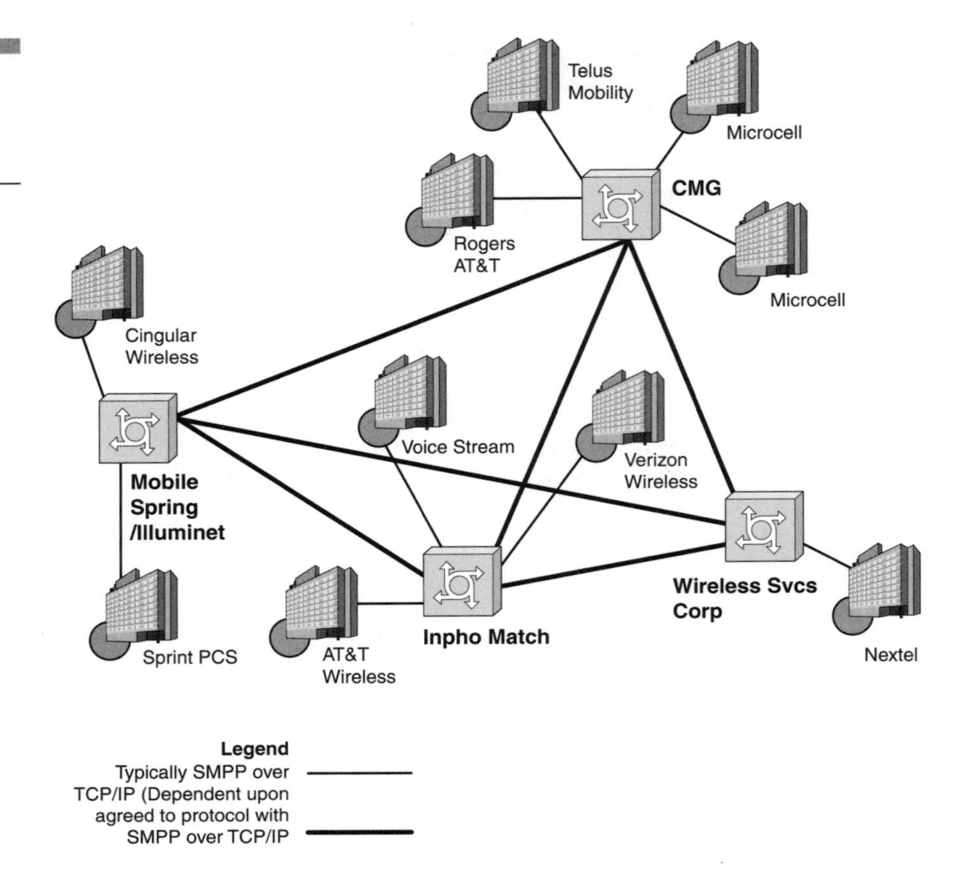

Figure 5-5 SMS interconnection (Source: CMG)

It can be expected that the other carriers will join the SMS interconnect network in the next few years, so the interconnect percentage will be 100 percent. Table 5-2 illustrates the current number of subscribers in North America.

From late 2001 to mid-2002, the North American carriers will see an upsurge of SMS activity, achieving 1 billion messages per month over all technologies by June 2002. The CTIA announced in early July 2002 that all of the six largest wireless carriers will participate in a program that enables a customer of one carrier to send a text message to a customer of another and predicted the technology would change the way America communicates.

Table 5-2
North American wireless subscribers

Wireless Carrier	Number of Subscribers
Verizon Wireless	34,081,900
Cingular Wireless	24,484,300
AT&T Wireless	19,042,800
Voicestream Wireless	7,786,200
SprintPCS	14,782,800
Nextel	8,700,000
Microcell	1,288,400
Rogers AT&T	3,135,600
Bell Mobility	3,333,200
Telus Mobility	2,504,300
Total	119,139,500 (83%)

Source: EMC

WAP: The Mobile Internet

Now that we have discussed the mechanics of 2G messaging, SMS, it is time to look at WAP. WAP was established as an attempt to make the Internet accessible from a handheld device. The mechanics and predecessor of WAP, *Handheld Device Markup Language* (HDML), will be discussed in the following section.

Although most modern phones are WAP enabled, only a small percentage of the subscribers actually use WAP. The reasons why WAP has not yet become a success and whether WAP will have a second chance will be discussed after that.

The Genesis of HDML/WML and WAP

HDML enables wireless access to a web site. Wireless access means that end users can access the web site from a cell phone, *personal digital assistant* (PDA), or other handheld device. Of course, wireless access of web sites through tiny screens places limitations on what can actually be displayed on the device. For this reason, web sites supporting HDML have tuned their web site navigation to the capabilities of the small handheld device. HDML, which is better than *Hypertext Markup Language* (HTML)—the standard Internet markup language—is capable of navigating a handheld device user through the web site by displaying only a limited amount of text at the same time. HDML is optimized for the usage of wireless request/wireless response application types, where the end user requests small bits of information at a time, such as airline information services, stock quote services, and so on.

HDML was invented in 1997 by Unwired Planet. Unwired Planet also invented a special browser for the HDML language, which it called the UP.Browser. The WAP standard was established later by the standardization forum named the WAP Forum, which was, not surprisingly, founded by Unwired Planet. Unwired Planet renamed itself to Phone.com and later to Openwave. Other active contributors to the WAP standard were the major cell phone manufacturers Nokia and Motorola.

WAP is supported by almost every phone vendor, and since it is an open standard, it has a large penetration rate throughout the world. This contrasts HDML, where HDML-capable phones are mostly used in North America.

Wireless Markup Language (WML) is part of the WAP standard. WML was derived from HDML, but since it was developed later than HDML, WML is more feature rich than HDML. Also, WML is fully compliant with *eXtended Markup Language* (XML), whereas HDML is not.

As can be derived from the previous statement, most modern phones support WML. For a wireless application provider, WML is therefore the most important markup language to use; however, for wireless application providers targeting the North American market, it is important to realize that most older phones support HDML, but do not support WML. A similar discussion of whether application providers should use HDML versus WML in the earlier discussion of choosing between SMPP versus UCP versus OIS. As a result, a wireless application provider often supports both WML and HDML, and has an automated detection mechanism built into the web site no matter if a WML or HDML browser is used.

The Mobile Internet Hype Debacle

WAP was positioned to make the Internet truly wireless. WAP was established in the dot-com era, during a time when most people had unrealistic expectations of Internet business. Combining the two buzzwords of that time, *Internet* and *wireless*, was a successful formula for creating hype. Stock quotes of companies that had anything to do with WAP went sky high.

Together with the dot-com hype, the WAP hype disappeared. The question is why did WAP fail? The following lists a number of factors that hampered the commercial success of WAP:

- *It is expensive*. Unlike Internet access by wireline, wireless access to the Internet is expensive. Because a connection is kept open during a WAP session and because the data transfer rate is mostly a maximum of 9.6 Kbps, WAP sessions tend to become rather expensive. Furthermore, the connection time has no relation to the value of the information that is accessed on the Web through a WAP session.
- *Because the connection is so expensive, wireless carriers are not capable of charging based on the content that is being accessed*. This hampers a good revenue-sharing model for wireless application providers. A good revenue-sharing model is based on the value of the information, not the connection time. For example, most men would be willing to pay more for a picture of their favorite Hollywood actress without clothes than for the same actress with her clothes. The time needed to download a picture of a nude actress is less than the time needed to download a picture of the same actress with her clothes on, so the nude picture would be cheaper. This obvious doesn't make commercial sense.
- *Access to information was slow*. Because wireless modem connections were mostly used for WAP sessions, a connection needed to be set up first before the web page could be displayed

on the phone. In many cases, it could take well over a minute to access a certain piece of information on the Web. This doesn't sound too bad, but keep in mind, because of the WAP hype, WAP users were expecting an Internet experience comparable to wireline Internet.

- *WAP connections were dropped due to less-than-optimal network coverage circumstances*. As described previously, most carriers used modem connections (or circuit-switched connections) to enable WAP sessions. Data connections, however, are much more vulnerable to network coverage circumstances. This often resulted in broken network connections, and the end user often had to undergo the frustration of having to wait another minute for the selected web page to be displayed again.
- *The first WAP phones were unstable and sometimes completely froze during usage*. Although PC users are familiar with the nuisance of having to reboot their PC from time to time, this phenomenon was new to a cell phone. This instability of the first WAP phones did not help the reputation of WAP.
- *WAP phones need to be configured correctly to work*. Trying the WAP functionality with wrong phone settings (WAP gateway address, username/password settings, and so on) caused many disappointed users to discard the WAP functionality on their phone after failing to get the WAP phone to work the first time.
- *Only a tiny subset of the Internet can be viewed via WAP*. Web sites need to be written in WML rather than in the Internet's standard HTML when accessed through WAP. Most web sites make extensive use of graphics, which, although supported in a limited way on WML-capable phones, is not very suitable for WAP-capable phones.

SMS and WAP: Similarities and Differences

Similarities

- Both SMS and WAP can be used to access information services from wireless devices.

- SMS can be used as a bearer for WAP; WAP push always generally uses SMS as a bearer.
- Both WAP and SMS can be used for wireless P2P communication. WAP provides access to web-based e-mail, which can be accessed by both the originator of the e-mail and the recipient of the e-mail.

Differences

- WAP phones access web sites on the Internet; SMS services are generally in the operator's private IP domain.
- WAP users need to pay for the connection time on most occasions, whereas SMS users only pay to send a message.[9]
- WAP is usually connection oriented (circuit switched), whereas SMS is connectionless (packet switched).
- WAP-based P2P communication does not allow for the *Calling Party Pays* (CPP) model since both the originator and the recipient need to set up a WAP session. Experience throughout the world has proven that networks using the CPP charging model are generally more successful with messaging than networks that charge both the originator and the recipient. As such, CPP is generally perceived as a prerequisite for messaging success. In case of WAP mail, however, the recipient needs to set up a (charged) WAP session to access new e-mail. To notify the receiver that a new e-mail has arrived, an SMS is sent to the recipient of the WAP mail.
- The WAP standard is compatible with all types of cellular networks, whereas the SMS service (if defined on that particular network type) is mostly incompatible with each other (see the section "Wireless Gateways").

To access a web site through a WAP phone, the end user clicks on one of the links on his or her homepage or directly enters the address

[9] In U.S. networks, cell phone users are generally charged for sending and receiving a message.

of the web site. Regardless how the web site is selected by the end user, a request from the WAP phone is sent as a *Uniform Resource Locator* (URL) through the network to the WAP gateway. The WAP gateway sits between the carrier's network and the Internet, and provides functionality such as billing, subscriber access control, encoding, decoding, and protocol translation. Figure 5-6 gives a schematic overview of a WAP gateway.

Architecture Picture of WAP

The following sections explain the protocol layers of a WAP gateway.

WAP Datagram Protocol (WDP) The *WAP Datagram Protocol* (WDP) layer is the transport layer that sends and receives messages via any available bearer network, including SMS, *Unstructured Supplementary Services Data* (USSD), *Circuit-Switched Data* (CSD),

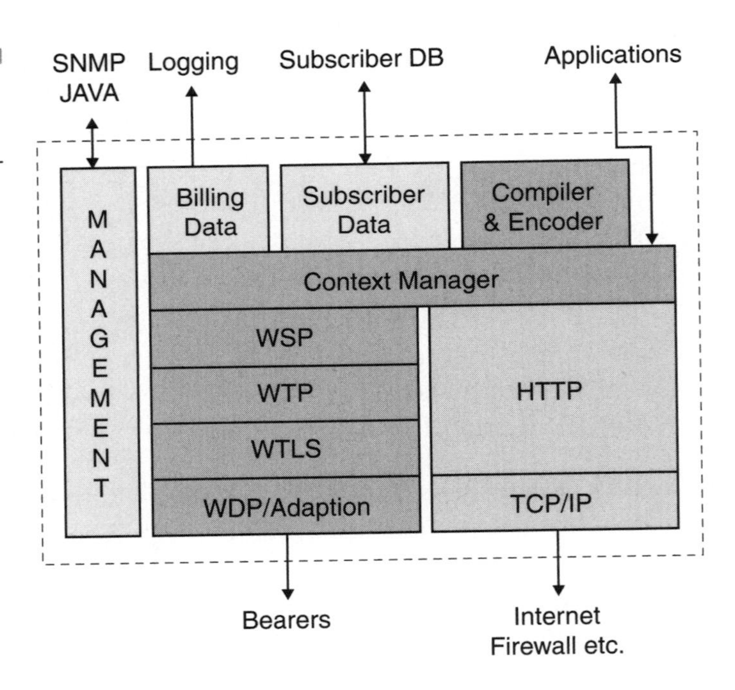

Figure 5-6 WAP gateway (Source: IEC.ORG)

Cellular Digital Packet Data (CDPD), IS-136 packet data, and *General Packet Radio Service* (GPRS).

Wireless Transport Layer Security (WTLS) WTLS, an optional security layer, has encryption facilities that provide the secure transport service required by many applications, such as e-commerce.

WAP Transaction Protocol (WTP) The *WAP Transaction Protocol* (WTP) layer provides transaction support, adding reliability to the datagram service provided by WDP.

WAP Session Protocol (WSP) The *WAP Session Protocol* (WSP) layer provides a lightweight session layer to allow the efficient exchange of data between applications.

HTTP Interface The HTTP interface serves to retrieve WAP content from the Internet requested by the mobile device. WAP content (WML and WMLScript) is converted into a compact binary form for transmission over the air.

Pros and Cons of SMS and WAP

Both technologies have their pros and cons. The preferred technology depends on the type of application and personal taste. Some pros and cons of SMS and WAP follow.

Pros of SMS

- The ability to receive SMS is available on 100 percent of the cell phones in GSM, CDMA, and TDMA. The ability to send SMS is available on nearly 100 in GSM and on the majority of the cell phones in CDMA and TDMA. This is a very strong argument for application providers to support SMS-based access to their services.
- SMS works under even the weakest network coverage circumstances.

- SMS can use a CPP charging model.
- SMS is instantaneous; there is no connection setup time.
- SMS is fast (quick delivery depending on SMSC used); it therefore enables a immediate dialogue between two people.
- SMS is suitable for interactive TV applications with a large audience (again subject to SMSC in question).
- SMS provides for value-based charging if SMS is used to access an application provider service.

Cons of SMS

- Some cell phones have a clumsy interface to access the SMS option.
- The phone displays *Message Sent* when in reality the message may be dropped if no SMS interconnect is available between the carriers of the originator and recipient.
- Worldwide SMS interconnect has not been established yet due to incompatibilities between GSM, CDMA, and TDMA.
- An application provider who wants to use SMS for end-user access to their services needs to make agreements with multiple carriers or alternatively needs the services of a gateway provider to arrange the connections to the individual carrier for them.
- For an application provider, providing services through SMS is rather expensive. This means that the business model of the application provider must allow for a considerable constant cost factor—the access to the carrier's network.
- SMS messages are limited in length. In GSM, SMS are limited to 160 characters. In TDMA and CDMA, the messages can theoretically be 256 characters, but in practice most are limited to between 120 and 140 characters because of network constraints.

Pros of WAP

- WAP enables a wireless end user to access information on the Internet.

- WAP technology allows for more user-friendliness of the application provider service.
- A security layer has been built into the WAP standard, making WAP suitable for financial transactions. For SMS to be suitable for financial transactions, SIM toolkit applications have to be downloaded first to the cell phone.
- A WAP application provider has low cost for being able to open up services through the network of a particular carrier compared to SMS.
- WAP mail offers worldwide connectivity through the Internet, regardless of the underlying radio technology.
- WAP mail can be delivered to any e-mail address.
- WAP mail messages can be longer in length than SMS messages.

Circuit Switched Versus Packet Switched

As mentioned earlier in this section, WAP can work over multiple so-called bearer channels. A bearer channel is a technology in a particular network type to move data from one side to another. The WDP layer adapts the WAP data to the bearer channel in use. Bearer channels can be circuit switched or packet switched.

Circuit Switched Circuit switched can be best compared to a voice call. A connection (circuit) has to be established through the network before data can be exchanged. The routing through the network is determined at the connection setup phase. After that, each piece of data will follow the same route as was determined at connection setup time; in other words, the data travels through the same MSCs and STPs.

A circuit-switched connection occupies a communication channel during the full duration of the WAP session. Using circuit-switched connections for WAP requires that for each parallel WAP session, a dial-in slot must be available; therefore, the number of parallel WAP sessions is limited. This situation is comparable to an *Internet service provider* (ISP) providing dial-up access to the Internet.

Circuit-switched WAP bearers are

- **CSD connections** As discussed previously, CSD connections are wireless modem connections through which data between the WAP gateway and the WAP phone is exchanged. As CSD makes use of a voice channel, CSD is available in GSM, CDMA, and TDMA networks.
- **USSD connections** USSD is a means of transmitting information or instructions over a GSM network. USSD is available on all GSM cell phones and can be accessed by the end user by typing in a number preceded by a * and ending with a #a and then pressing the phone call button. For example, *100#a sets up a connection to USSD application 100. USSD has some similarities to SMS since both USSD and SMS use the GSM network's signaling path. With USSD, however, when a user accesses a USSD service, the radio connection is kept open until the connection is terminated by either the user or by the application.

  USSD stage 2 is a standardized WAP bearer channel, but it is not so widely used as CSD. USSD connections do have certain advantages over CSD connections, however, because connection setup times are much faster with USSD than with CSD so the end user can access the information faster. Secondly, USSD connections are not as vulnerable for bad network coverage conditions as CSD connections.

Packet Switched Unlike circuit-switched connections, packet-switched networks do not need to set up a connection before data can be exchanged between two end points (in this case, the WAP phone and WAP gateway). This is possible because data is encapsulated in data packets, where each packet has header information containing the sender and the destination of that particular packet. Based on the header information, the network can decide where that packet should be routed. The routing of the data is determined dynamically on a per-packet basis—hence the term *packet switched*. For example, the Internet is a packet-switched network because the Internet's IP packets can be routed dynamically through the Internet backbone.

The charging of packet-switched connections is usually based on the amount of data (kilobytes) exchanged during the data communication session.

Packet-switched WAP bearers are

- **SMS** Although not optimal, SMS is available as a packet-switched bearer channel for WAP. Not many carriers use SMS as a WAP bearer channel; nevertheless, it does have some advantages over CSD and USSD:
  - SMS provides WAP access under even the worst network coverage circumstances.
  - For simple request/response WAP services, SMS is faster than CSD (but not faster than USSD).
  - SMS does not occupy network resources during the WAP session, which is the case for CSD (voice channel is occupied during WAP session) and USSD (signaling channel is occupied during WAP session).
  - There is no limitation to the parallel number of WAP sessions to the WAP gateway if SMS is used as a bearer. This is an issue with CSD and USSD.
- **CDPD** CDPD offers wireless packet-switched IP access to handheld devices over TDMA and *Advanced Mobile Phone System* (AMPS) networks up to theoretically 19.2 Kbps. Since CDPD does not offer voice capabilities, its usage is optimized for PDAs and the like. CDPD is especially popular in North America, where most major cities have CDPD coverage.
- ***Generic UDP Transport Service* (GUTS)** GUTS has been defined as a similar service to SMS in a TDMA network for transmitting UDP data packets from an application to a cell phone, and vice versa. So far no commercial WAP phones have been brought to market that support GUTS as a WAP bearer channel.
- **GPRS** Similar to CDPD, GPRS offers wireless packet-switched IP access to handheld devices. GPRS is defined as an add-on to GSM and can theoretically provide bandwidth up to 34 Kbps.

The Future of WAP and WML

Although WAP was perceived as a promising technology at the time of market launch, a few years later the public opinion is that WAP has failed. However, looking at the reasons for the failure of WAP described in the section "The Mobile Internet Hype Debacle," the failure had little to do with the WAP service itself; it was more because of the factors around it, such as an expensive bearer channel and billing facilities. The conclusion from this is that the failure of WAP was caused because WAP was launched too early since the necessary technologies were not ready yet.

i-mode In Japan, the mobile Internet is immensely popular with more than 10 million users using a service called *i-mode*. i-mode is owned by the Japanese carrier NTT DoCoMo. i-mode differs from WAP in a number of ways:

- i-mode provides fast access to information by offering packet-switched access to the network.
- i-mode uses a subset of HTML called *Compact HTML* (cHTML), which makes it easier for web application developers to adjust their web service for i-mode.
- i-mode includes a business model where both the web application provider and the carrier make a profitable business.

However, compared to WAP, i-mode has a weak security model, which is why NTT DoCoMo has been looking to extend i-mode with security features.

WAP 2.0 and XHTML Recently, the WAP Forum released a new version of the WAP protocol: WAP 2.0. With version 2.0, WAP will move toward more accepted Internet standards such as *eXtended Hypertext Markup Language* (X-HTML) and TCP. As a result, it is expected that more WAP applications will become available, as with X-HTML—similar to i-mode's cHTML—only little adaptations are necessary for a web site to be viewed through WAP 2.0 phones.

In previous versions of WAP, a WAP gateway was required to handle the protocol interworking between the WAP phone and the web server. The WAP gateway communicated with the WAP phone using the WAP protocols that were largely based on Internet communication protocols, and it communicated with the web server using the standard Internet protocols. In principle, WAP 2.0 does not require a WAP gateway since the communication between the WAP phone and web server can be conducted using end-to-end HTTP. However, a WAP gateway can optimize the communications process and may offer mobile service enhancements, such as location-, privacy-, and presence-based services. In addition, a push proxy is necessary to offer push functionality.

While the WAP Forum is moving WAP toward XHTML and TCP, NTT DoCoMo's i-mode is moving in the same direction to improve the security features. WAP 2.0 and i-mode are expected to merge somewhere in the near future.

WAP 2.0 makes use of a wide range of new technologies and advanced capabilities, such as

- **Networks and network bearers** Carriers worldwide are upgrading their existing networks with higher-speed bearers such as GPRS and *High-Speed Circuit-Switched Data* (HSCSD) and introducing higher bandwidths and speeds in *third-generation* (3G) wireless networks such as *Wireless CDMA* (W-CDMA) and CDMA2000 3XRTT. These more capable network bearers permit new types of content (such as streaming media) and provide an always-on availability. These new aspects of the serving networks permit new operational activities.
- **TCP/IP as Transport Protocol** Most new wireless network technologies provide IP packet support as a basic data transport protocol. WAP 2.0 uses a mobile profile of TCP for wireless links. This profile is fully interoperable with the common TCP that operates over the Internet today.

In combination with faster, packet-switched bearer channels such as GPRS, WAP 2.0 will offer a far better end-user experience and has a good chance of becoming the popular service WAP was expected to become back in 1999 and 2000.

Pull Versus Push The WAP pull service is the basic WAP service; a WAP user enters a URL of a web service either directly or through a menu choice. The URL request is sent through the WAP gateway onto the Internet and subsequently the WAP page is pulled off the Internet.

The WAP push service enables content to be sent or pushed to WAP phones by web applications via a device called a *push proxy*. Push functionality is especially relevant to web applications that send notifications to their users, such as messaging, stock price, and traffic update alerts. Without push functionality, these types of applications would require the WAP phones to poll the web application for new information or status. In wireless environments, polling activities would become expensive for the end user and would mean a wasteful use of the resources of wireless networks.

If GPRS or CSD is used, and no WAP session is in progress, WAP push messages use SMS as the underlying bearer channel. If a WAP session is in progress, the WAP push message is sent through the active connection (that is, over either the GPRS network or the CSD connection). In 3G networks, it is possible to send WAP push messages by having the push proxy initiate a TCP/IP connection to the cell phone.

CHAPTER 6

Wireless Messaging's Future: A Look Down the Road

The explosive growth of wireless messaging since its introduction in the United Kingdom, continental Europe, and Asia about a decade ago promises optimistic prospects for the medium. In 2001, 700 million mobile phone users worldwide sent an average of 20 billion text messages a month, generating billions of dollars of revenue for wireless carriers. Only one year later, that number has ballooned to an estimated 30 billion messages per month. Figure 6-1 demonstrates the steep growth curve of global *Short Message Service* (SMS) average monthly messages from early 2000 to the end of 2001.

As the advanced features of *Enhanced Messaging Service* (EMS) become more widely accessible in parts of the world and *Multimedia Messaging Service* (MMS) becomes available, an exponential increase can realistically be expected as the intended target market expands from the youth demographic, which catapulted SMS into the global consciousness, to the business and entrepreneurial market.

SMS is already growing beyond the youth market to include countless practical applications that can be exploited by diverse groups. Text messaging has become increasingly popular among the deaf population in Europe, for instance, because it has opened new lines of communication. Businesspeople find text messaging both

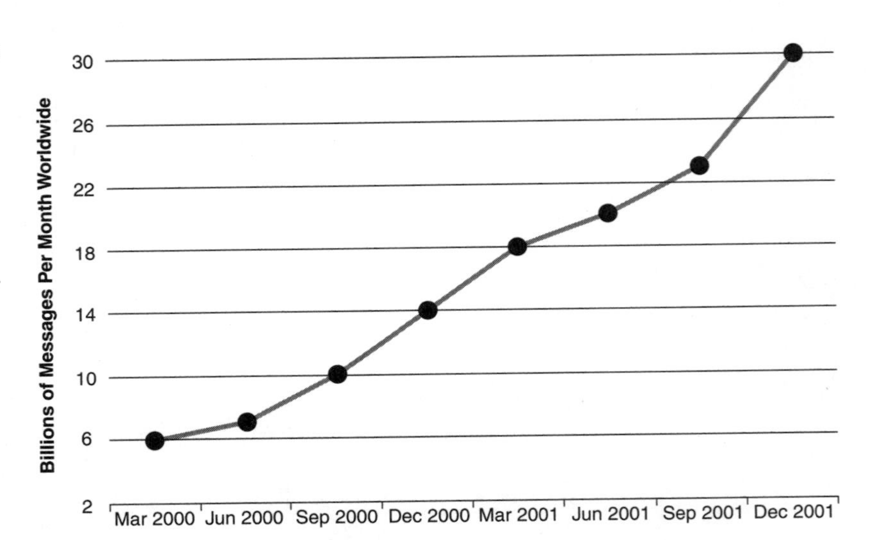

Figure 6-1 Global SMS growth from March 2000 to December 2001 (in billions) (Source: Arena Intelligence Group)

convenient and discrete for exchanging information anytime and anywhere. Parents use text messaging to tactfully keep tabs on their children's plans and whereabouts. Enterprises also use it for myriad purposes, from sending alerts, or reminders, to customers to processing ticket sales.

Person-to-person (P2P) messaging accounts for 90 percent of global message traffic. The other 10% is made up of information services (such as news, sports scores, weather reports, public transportation schedules, and stock quotes) and alerts (largely in the banking and financial services industry). Imagine the advantage of receiving a text message when your checking account balance is dwindling. You would not have to worry about accidental overdrafts.

SMS is also gathering steam as a medium for television audience voting and as a means of canvassing mass audience opinion for televised awards ceremonies. Hugely successful European programs, such as *TXT-REQUEST*, *Who Wants to Be a Millionaire*, *Survivor*, *Big Brother*, *Pop Idol*, *Popstarz*, and the BBC's "I Want Your Text" radio competition are just the beginning to fully exploit this massive revenue opportunity. Bolstered by the success of interactive television programs, SMS payment at vending machines, where users can purchase snacks and soft drinks by keying a password into their mobile phone, is already a reality in some countries (for example, Scandinavia). Other principal uses for SMS include *Unified Messaging* (UM), chat, job dispatch, customer service, affinity programs, and remote monitoring.

However, with any new technology, fresh challenges arise. The new-generation networks—*2.5 generation* (2.5G) and *third generation* (3G)—required to provide many of the services promised are expensive and behind schedule, and have been met with lukewarm acceptance. Although their implementation is incomplete, 2.5G and 3G services have sometimes failed to live up to the expectations of consumers who envisioned portable devices with capabilities far beyond those currently offered. At the same time, enormous potential exists, despite the fact that in North America these devices have only evolved slightly from the current *second-generation* (2G) devices. This disappointment has created speculation about the next generation: *fourth generation* (4G). Although 4G is essentially just conceptual guidelines at this point, research and development is

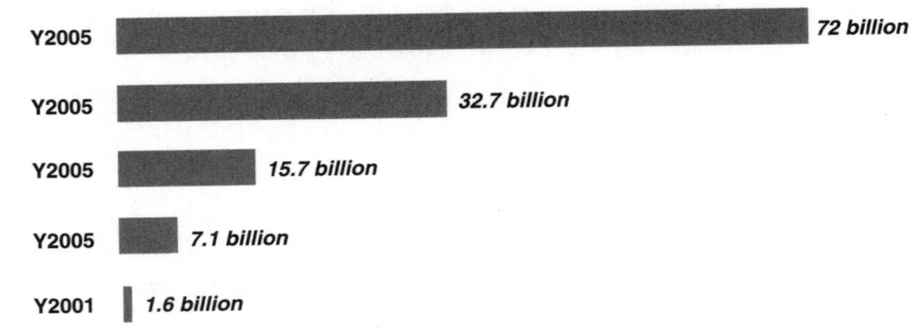

Figure 6-2 U.S. SMS message growth from 2001 to 2005 (from mobile-initiated SMS) (Source: Arena Intelligence Group)

moving ahead. 4G might become a reality as early as 2006—four years earlier than the original rollout target date.

Gartner, Inc., a leading technology research and advisory firm, indicates that "as many as 800 million consumers worldwide will use wireless data services by 2004," and three years later "more than 60 percent of 15- to 50-year-olds living in the United States and Europe will carry or wear a mobile device."[1] Increasingly, teenagers acknowledge that they send messages until their thumb aches. Wireless industry analysts' projections for SMS message growth in the United States, as shown in Figure 6-2, lend credence to such a positive scenario for user adoption.

Indeed, wireless messaging is here to stay, and its applications are limited only by the imagination. Wireless messaging holds something for everyone—and there's more to come.

Wireless Messaging—Post 2G

Globally, SMS has become the fastest growing data service in the wireless industry. Both individuals and businesses are increasingly adopting messaging solutions. Consequently, wireless data services

[1]Kristy Bassuener. "Wireless Will Win Consumers Over." *Cahners Wireless Week Online*. Erols. (May 8, 2001).

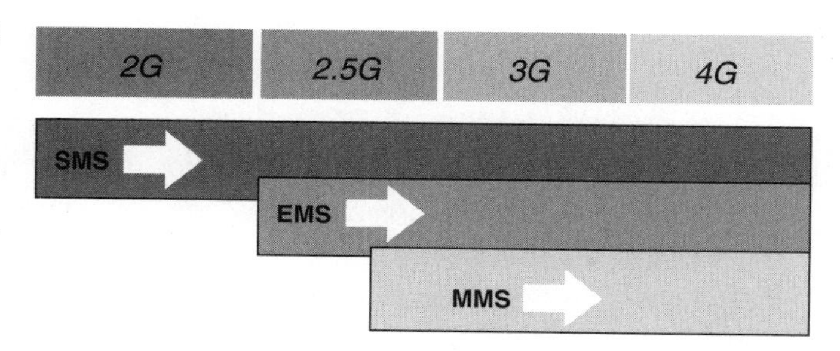

Figure 6-3 Evolution of wireless messaging (Source: Arena Intelligence Group)

represent approximately 15 percent of revenues for European, at most, 35% for Asian wireless carriers, generating billions of dollars annually in SMS revenue. Heeding the example of their overseas competitors, North American and Latin American carriers are starting to develop profitable new revenue streams by offering wireless data services to a large, but underserved, market of millions of mobile phone users. Figure 6-3 illustrates the technologies represented throughout the evolution of the wireless messaging services.

The progression from the current 2G networks to the next-generation 2.5G, 3G, and 4G networks will greatly influence these data services in many different ways. Most notably these changes will affect the following:

- **The message** The most obvious change involves the messages themselves. As the networks gain speed and the amount of data that can be transferred increases, the once simple SMS text message will become more complex. Known as EMS, the early enhancements have taken the form of ring tones, sound effects, and black-and-white pictures (such as logos and wallpapers). Eventually, as the 3G and 4G networks roll out, the next-generation message formats will incorporate multimedia aspects such as audio, video, and file downloads.
- **The medium** Another notable but less obvious change concerns the method by which these messages are delivered. Even though EMS messages are sent through the same network structure as SMS messages, MMS will require the full implementation of 3G's bandwidth potential to function and 4G's faster-than-DSL speed to bring about the promised media-rich content.

- **The market** The third major change relates to the nature of the consumers of these services. Right now, the predominant users of SMS worldwide are young people in their mid-teens to their mid-twenties. The youth market quickly adopted the new technology for several reasons. When 520 students from 10 British universities were surveyed, 50 percent reported that their monthly mobile phone bill was £20 ($33) and that they preferred text messaging for its cost effectiveness. Convenience for communicating with their friends was given as another motivating factor for SMS use. More than 75 percent of those students said they send text messages everyday, and more than 33 percent said they send over five messages a day.[2]

Affordability and convenience notwithstanding, a less quantifiable aspect has also helped to drive the success of SMS among young adults. Trendiness, although difficult to measure, has undoubtedly influenced the rapid adoption of SMS. Style is paramount in the teenage psyche worldwide, and text messaging has become as important to peer acceptance as the latest fashions in clothing. Pithy and code-like in its shorthand construction, the language of SMS appeals to young users, drawing them into the circle of the technology savvy.

Although some services employ SMS to convey information other than simple text messages (such as weather, sports scores, stock market quotes, and ticket services), most SMS messages constitute private communications exchanged between individuals. This is significant because new EMS and MMS services will probably be heavily used for personal messaging, but will not be as heavily used as traditional and less expensive SMS.

In order for the next generation of wireless messaging services to succeed, it will have to attract an older, more professional demographic because EMS and MMS will cost more than SMS—perhaps 10 times more at the outset. Providers of 3G services will attempt to attract this market segment by offering expanded versions of services that have already established popularity in their SMS version, such as interactive banking and real-time chat.

[2] "Text Messaging Very Popular Among Students." *Ananova* Online. Erols. (January 24, 2002).

3G Technology

3G, which is actually just a general term referring to a number of protocols and standards that will be used in the next generation of wireless services. Please refer back to Chapter 2, "Technology and Market Overview," for 3G and technology discussion.

3G's different protocols and standards are spread across the three main global markets. Needless to say, they hold the potential to cause havoc in standardization and cross-provider interoperability. If these elements obstructed the exchange of information between different service providers, 3G would have little hope of ever getting off the ground or gaining consumer support. The *Third-Generation Partnership Project* (3GPP) and the *Third-Generation Partnership Project 2* (3GPP2) stepped into this potential morass.

Third-Generation Partnership Project (3GPP) The 3GPP was formed at the suggestion of ETSI in December 1998. The original mandate was to develop global technical specifications and technical reports for a 3G wireless system based on the *Global System for Mobile* (GSM) communication network core and the radio access technologies it supports. Later, this task was amended to include the maintenance and development of the GSM technical specifications and technical reports, including evolved radio-access technologies (such as GPRS and EDGE). *Release 99*, a fundamental step in the evolution of 3G technology (the definitive 3G standard), was completed in December 1999.

Release 99 enabled developers to move ahead with the design of a standards-compliant infrastructure and terminal. Subsequent changes to Release 99 were codified in Release 4 in 2001, and Release 5 was introduced in 2002, laying the groundwork for integration of 3G networks and *Internet Protocol* (IP) technology. This confirms the increased 3G product development by equipment vendors. This codification of the network protocols will enable rapid service growth once 3G is totally online.

Third-Generation Partnership Project 2 (3GPP2) While ETSI's 3GPP focused on standardizing GSM networks and the telecommunication technology they support, another partnership involving Asian and North American interests was formed to cover the issues involved in setting global standards for the migration of ANSI-41's *radio transmission technologies* (RTT) to 3G. This parallel partnership, called 3GPP2, benefits from the collaborative effort and the recognition it has gained as a specifications-setting endeavor.

In the early discussions of 3GPP's formation, it was suggested that ETSI and the ANSI-41 communities join forces. Later, they realized that parallel partnerships would help the eventual standardization to proceed quicker within their own markets and then merge at some point in the future. Plans for the consolidation of the two partnerships were built into the structure of the organizations themselves. Each organization invited observers from a different organization.

Next-Generation Wireless Messaging Services

The application's relatively low cost, high message volume, and technical simplicity have made SMS technology a rich revenue source for wireless carriers. Motivated by likely revenue increases if more features were added to the simple text message, handset manufacturers and carriers have acted on a perceived need for improved services. By using the higher bandwidth and faster download speed inherent in 2.5G and 3G, carriers gain the potential to turn the next generation of messaging services into a more stylized and personal experience. Images, sounds, and animation are now available. Full multimedia messaging features will be available as soon as the 3G and 4G networks have been completely implemented.

Because most SMS messages occur between individuals, many added features of next-generation technology serve to make the message more personalized, fun, whimsical, or noticeable. Wireless-message-based businesses can also use these features. In fact, many of the new features that make a friend's message fun to read (such as images and ring tones) also increase the likelihood that a recipient will read a business or advertising message.

Table 6-1

Wireless messaging technology capabilities

SMS	EMS	MMS
Standard text	Standard and formatted text Black-and-white pixel images Simple animations Sound Ring tones	Standard and formatted text High-quality images Video Melodies and other audio

Source: Infomatch

Table 6-1 differentiates the capabilities of wireless technology through the progression from 2G SMS to 2.5G EMS/MMS to 3G–4G MMS.

EMS

EMS is the next step in the evolutionary process after SMS and has been available since 2001 in some parts of the world. EMS enables users to send and receive a combination of pictures, animations, sounds, simple melodies, formatted text, and standard text as an integrated message between EMS-enabled mobile handsets. The technological standard, which was developed by the 3GPP, adds innovative features to standard SMS messaging:

- **Images** Users can display images in three sizes: small (8 × 8 pixels), large (16 × 16 pixels), and variable, depending on the format the provider supports.
- **Animations** Users can select predefined animated images (similar to emoticons). These reside in the handset or other wireless device and can be displayed in two sizes: small (8 × 8 pixels) and large (16 × 16 pixels). Animations cannot actually be sent because the SMS and EMS infrastructures can convey messages only in small bandwidth slices that are unable to handle the large amount of data needed.
- **Sounds** Users can also add predefined sounds to a message. Like animations, most sounds reside in the receiving device and

are activated by commands within the incoming message, although new sounds can be attached to an EMS message.

- **Text formatting** Users can employ basic text formatting within messages. Italics, bolding, underlining, justifications, and fonts add to the personalization unavailable with SMS.
- **Larger messages** Will be sent through multiple SMS messages.

One of the greatest advantages about EMS is its capability to use the existing SMS infrastructure. As a result, carriers do not have to invest in new networks and the service can be easily adopted and deployed. In addition, since there is no requirement for new technology, the per-message price remains low. EMS also provides a certain amount of backwards compatibility by enabling the text portions of an EMS message to be received on a device that is only SMS enabled. Moreover, because the EMS source code is open, any handset manufacturer may develop applications.

Fortunately, challenges to EMS are minor. In a few instances, compatibility problems occur when users attempt to exchange EMS messages between handsets from different manufacturers, even though both have adopted the EMS standard. By taking a wait-and-see stance, several handset manufacturers have been slow to deploy EMS. Nokia has chosen not to adopt it at all; instead, it employs a proprietary technology (usually incompatible with EMS) to send enhanced messages, known as Smart Messaging.

Overall, however, EMS provides features at an affordable per-message price that P2P users enjoy and vendor-to-consumer users feel they need. Again, wireless industry experts predict that the youth market that boosted SMS to great heights will lead the way for EMS.

EMS Mechanisms

EMS was initially designed to enable picture messaging and the exchange of ring tones for all handsets. EMS adds a number of encoding mechanisms to existing SMS messages to enable richer content than just plaintext. Note, however, that the SMS transport and delivery mechanisms are untouched when introducing EMS.

The EMS extension is actually based on the way messages are coded. A normal SMS message contains a header followed by the actual content of the message, which has a maximum length of 160 characters. A part of the header, called the *User Data Header*, allows the introduction of binary data within an SMS message. When an EMS message is created, this header always indicates the presence of binary data for formatting the text and images within the message.

The use of SMS message concatenation overcomes the second problem of EMS. Standard messages are bound to 140 bytes (or 160 characters), which is too small for pictures and animations. SMS has a mechanism of coupling several 140-byte messages together into one single logical message. The maximum number of concatenated messages is 255, which leads to a maximum EMS message size of 38KB.

The ability to add binary content to messages together with the ability to create large messages lay the foundation of EMS. The next section will detail the possible contents of an EMS message.

Capabilities in EMS Messages

The EMS specification actually permits a range of additions to simple text messages. The main enhancements of EMS include the following:

- Text formatting
- Small and large pictures
- Sounds and standard sound effects
- Animations
- *Wireless Vector Graphics* (WVG), which can be used for drawings and handwritten symbols, such as Kanji (Japanese character set)

Text Formatting EMS defines a number of parameters for the formatting of plaintext within an SMS message. With EMS, the user can influence text alignment (left, right, and center), font size (small, normal, and large), and font style (bold, italics, underline, and strikethrough). This can result in the message shown in Figure 6-4.

For the newest devices with color screens, even background color and font color can be manipulated.

Figure 6-4 Example of EMS message (Source: CMG)

This is **not**
just simple SMS
message.

This is **EMS**

Joe

Sound Formatting The EMS specification differentiates between predefined sounds and user-defined sounds. Predefined sounds are efficient since they do not have to be transported over an over-the-air interface. However, the presentations of these sounds are (handset) vendor specific. EMS identifies the following predefined sounds: chimes high, chimes low, ding, ta-da, notify, drum, claps, fanfare, chord high, and chord low.

Enhanced messages can also contain user-defined audio clips in the iMelody format, which has a maximum length of 128 bytes. iMelody is a monophonic sound format that is mainly used for ring tones. Therefore, most EMS devices contain an iMelody composer. However, content providers provide most tunes.

Animations An enhanced message has two possibilities for animations. While creating an EMS message, users can include a predefined animation—for example, flirty, happy, sad, winking. Fifteen predefined animations are available.

User-defined animations can also be included in an EMS message. A user-defined animation consists of four pictures (8×8 and 16×16 black-and-white pixels) displayed sequentially. Since it is difficult to create such animation on the mobile device, this capability is defined mainly for *mobile-terminated* (MT) services.

Wireless Vector Graphics (WVG) The latest additions in the EMS specification include vector graphics. These are scalable pictures or animations that can be included in a message. A special distinction of vector graphic is the so-called character-size WVG element, which is wrapped in the SMS text as a handwritten char-

Figure 6-5 Example of WVG element (Source: CMG)

Joe, I ♡ you!

acter, for example. This can result in the EMS message shown in Figure 6-5.

MMS

MMS is less textual and more visual than EMS. It enables users with MMS-enabled handsets to exchange messages that contain text, sound, pictures, and video. MMS is a giant step in the wireless messaging evolutionary process. The possibilities for MMS stagger the imagination, leading large industry players to predict that the potential revenue from MMS could become the principal factor in recouping the vast amounts of money invested in 3G networks.

MMS—Dial *M* for Multimedia

While EMS enhances the SMS text service by introducing formatting and elementary bitmaps, true multimedia messaging has arrived with the launch of MMS. With MMS, users can exchange high-quality pictures (JPEG and GIF formats) and full-color animations (GIF format), sound clips (MIDI polyphonic tunes, MP3 music clips, and AMR voice recordings), and, in the near future, video clips (MPEG-4); these are all well-known formats in the PC and Internet environment.

The exchange of multimedia messages can lead to an explosion in message size—the initial service is launched with an average message size of 30KB. This should not be taken lightly when building the infrastructure behind the service. MMS also puts harsh requirements on the handsets both in creating and rendering multimedia content. From a user perspective, MMS may seem like a small step from EMS, but technology wise, it is a quantum leap. We will explore this topic in the following sections.

In order to gain acceptance and increase the consumer's willingness to pay, MMS must provide what it promises: full multimedia. In other words, MMS must combine text, audio, images, and video into a single message that can be delivered quickly and efficiently. Proponents of MMS guarantee a messaging service that includes the following:

- Standard audio formats such as MP3 and *Multipurpose Internet Mail Extensions* (MIDI)
- Standard image formats such as *Joint Photographic Experts Group* (JPEG) and *Graphics Interchange Format* (GIF)
- Standard video formats such as *Moving Pictures Experts Group 4* (MPEG-4)
- Migration of PC utilities (such as the personalization of desktops, screensavers, and plug-ins) to wireless devices
- Close interworking with the Internet using standards such as *Simple Mail Transfer Protocol* (SMTP) for e-mail and MIME for transporting multimedia files

In fact, this last point represents a critical element of MMS as the line between the Internet and wireless devices becomes less distinct. For example, an MMS message can be sent between MMS devices, to an e-mail address, or from an e-mail address to an MMS-enabled handset. Also, if an MMS message is sent to a non-MMS-enabled device, an SMS message will direct the recipient to a web site where the full MMS message can be retrieved. The use of the already-familiar Internet and PC technology will encourage consumer receptivity to similar services offered on their wireless devices.

The new MMS-enabled devices embody an entirely new generation of wireless technology. Moreover, the high bandwidth that MMS's media-rich content demands will necessitate the full implementation of 2.5G and 3G wireless networks. Should these bandwidths fail to materialize, full MMS, such as streaming video, might have to wait until the arrival of 4G networks.

Along with being exciting, fun, and useful, the new services and enhancements of MMS are expensive to implement, representing a major investment for wireless carriers who want to offer them to their subscribers. Unlike SMS and EMS, which could be slipped into available wireless network voice channels and thus enhanced revenues for carriers, MMS will require an entirely new infrastructure, as previously discussed.

The considerable financial outlay necessary to ensure the viability of MMS has slowed adoption by carriers, despite their initial eagerness. Even though wireless industry experts correctly predicted the emergence of MMS in 2002, only the most cutting-edge 2.5G and 3G carriers have offered it in limited areas.

The 3GPP is currently setting the standards for 3G. The 3G networks represent the ultimate future of MMS. However, MMS is offered on GPRS (or the so-called 2.5G) networks. (To support MMS, the existing GSM networks need a *Multimedia Messaging Service Center* [MMSC], which is a gateway of sorts to transfer between the networks. This involves an added expense and is another reason to push for the full implementation of 2.5G and 3G networks.) Although the GPRS system limits some of the multimedia features such as video (which is currently a series of slides similar to presentation software), it enables users to become acquainted with the technology. When the 3G networks are completely developed, the enhanced features of MMS will come to the fore, and the video and audio will move to the promised streaming versions.

If the delays of 2.5G and 3G implementation have dampened enthusiasm for MMS in the short run, wireless industry observers harbor little doubt that adoption will occur. Indeed, industry support behind MMS indicates a bright future, even though it could take up to 10 years for MMS to attain the 80 percent user adoption level that SMS currently enjoys.

MMS Around the World Most western European wireless carriers have MMS plans in various stages of development. This region is generally regarded as a hotbed of MMS development where many of the early MMS services will begin. More than 20 providers have announced MMS launch plans. Norway's Telenor was the first to introduce a commercial service.

The adoption of MMS will vary throughout Asia, although the region is viewed as rich in possibilities. Providers in Hong Kong have several MMS-based systems up and running, Singapore's M1 is expected to launch service by the end of 2002, and China has already started trials of micropayment MMS services.

The clearest example of how MMS will look and function, however, exists in Japan, where J-Phone has launched MMS service. The service offers impressive voice quality as well as mobile Internet service, video-clip distribution, and videophone service.

North America will trail Asia and Europe in MMS deployment by about 18 months. Once again, this lag has been caused by the delay in 2.5G and 3G deployment and the consumer market's lukewarm reception to the new services.

SMIL—Got the Message?

A first difference with EMS is the message itself. An MMS message is a MIME-formatted message; MIME is an Internet standard for message formats and is also used by e-mail services. MIME messages can contain one or multiple MIME types (for example, image, audio, or text). Typically, an MMS message also includes a presentation part, using a subset of the SMIL 2.0 specification. SMIL is an XML-like scripting language that is used to relate different MIME types together into one presentation. Using the SMIL presentation, multipart messages are played to the user so that it becomes an animated presentation. A message is split into multiple pages, which each have a timer. Each page on its own may contain one image, one audio clip, and one text part.

One of the benefits of using the MIME format is that it is scalable—it does not limit the message size in any way. In addition, it makes messages interoperable with the Internet environment. To ensure interoperability between handsets, however, the initial message size needs to be between 30 and 75KB.

Currently, an MMS message can contain images (JPEG and GIF), animations (GIF), audio clips (AMR, MIDI, MP3, and iMelody), text (in all known character sets, for example, ASCII, Arabic, and Chinese, which are all encoded in the Unicode standard), and vCards. In a later stage, video formats may be added (H263 format).

A second significant difference between MMS and SMS becomes clear when sending a message. The MMS message has a header that closely resembles an e-mail header and also facilitates the same type of service options, such as the request for a delivery report, the ability to address multiple recipients at the same time, message priority settings, and the request for a read reply. When addressing message recipients, both mobile phone numbers and e-mail addresses can be used.

The Future of MMS

With the commercial launch of MMS in early 2002, a new era in mobile messaging has started. However, there is still a long way to go. MMSC interconnection still has to be arranged, and interoperability between handsets and infrastructure has not reached a steady state yet. Despite the short-term challenges, the main MMS players are looking farther down the road. The addition of video content is only a matter of time, and the use of streaming technology is already facilitated within SMIL. The SMIL specification enables the presence of RTSP session setup within a SMIL presentation. The pace of innovation in the video area within MMS is actually determined by the availability of more powerful handsets.

MMS has some serious bottlenecks that need to be solved before the service can actually be as successful as some industry analysts predict. For content providers, *digital rights management* (DRM) is key for MMS to be an attractive investment area. As long as the illegal copying and forwarding of premium content is possible, content providers might stay clear of distributing multimedia content to mobile phones. In addition, the legislator will also monitor the use of MMS service closely since facilitating distribution of illegal content is considered a serious crime in some countries.

However, despite the challenges, many see MMS, the multimedia successor of SMS, as the prelude of a new era in mobile services. Multimedia services must flourish to provide the necessary *return on investments* (ROIs) made by the operator community in next-generation networks. Therefore, operators have a vested interest in continuing the messaging wave and want MMS to have the same returns they have with SMS.

What happens if my wireless service provider does not offer enhanced messaging? Can I change wireless service providers and take my current mobile number with me and use the messaging services?

Wireless Application Protocol (WAP) and the Wireless Internet Reborn

A consortium of companies that includes Openwave Systems, Ericsson, Motorola, and Nokia created the *Wireless Application Protocol* (WAP), which is also referred to as the *wireless Internet*. This group set the standards for a mini-web-browser that has given portable devices, such as mobile phones and *personal digital assistants* (PDAs), access to the Internet.

A wireless device can access a web site through the mini-web-browser and convert the *Hypertext Markup Language* (HTML) data into a form of markup language that the device can read, in most cases, as text. This markup language is called *Wireless Markup Language* (WML). The small screen size and limited download speed prohibit a true alternative to the full Internet, but they provide access to the Internet for e-mail and other text-based services for users who are away from their traditional Internet interface.

The Future of WAP The emergence of next generation wireless networks and their potential high-speed capabilities *might* render the WAP technology obsolete since it was developed for use with the low bandwidths available on current wireless networks. For this reason, the WAP consortium has pushed for a new generation of WAP standards that will function with the new wireless networks and finally bring the concept of wireless Internet to fruition.

The 3G partnerships are monitoring and supporting the next generation of WAP standards and devices, referred to as *WAP-NG*, as a possible standard wireless web browser for the new 3G handsets. NTT DoCoMo has also become heavily involved in this development to find a solution it can use with its i-mode service (i-mode currently uses a proprietary HTML-derivative application for its service). This new support from all the major 3G markets—North America, Europe, and Asia—is breathing life into a fully interactive wireless Internet that resembles the Internet that users are accustomed.

Wireless Messaging and Location-Based Services (LBS)

P2P messaging has become especially popular with mobile phone users so, naturally, vendor-to-consumer applications have been sought to capitalize on this popularity. Services that deliver weather, financial, and traffic reports represent several personalized applications that are making commercial use of SMS technology. The next step in the progression of commercial uses for SMS is *location-enabling* the service so information providers actually know the message recipient's location.

Location-Enabling Services

Through location-enabling, wireless carriers can offer their subscribers a multitude of new data services such as navigation assistance ("I'm lost. How do I get to Fifth Avenue and Main Street?"), concierge services, proximity searches ("What restaurants are nearby?"), and much more. These types of services benefit the frequent traveler, but can also be valuable for tourists and, on a more local level, for anyone who just happens to be in an unfamiliar part of his or her hometown. In addition, enterprises that employ a large sales staff or numerous field service personnel and businesses that rely on a highly mobile workforce to provide transportation, shipping, or delivery services are perfect candidates for location-enabling services. As an employee's location changes, the information that person receives changes, ensuring its relevance.

Information Delivery Services

As the wireless networks advance and carriers activate more 3G and 4G features and services, the type of personalized data that can be gathered from mobile phone users goes well beyond just location-based information. Advance *location-based services* (LBSs) is a step farther by taking advantage of the wireless device user's context and habits. This knowledge is then exploited by tailoring the information conveyed to that user. Applications draw on all the data that can be gathered about a user's situation based on the following:

- **Location** Location currently available to wireless providers
- **Presence** User location in relation to a network-enabled device (for example, at the computer, away from the desk, with mobile phone on, and so on)
- **Appointment calendars** Information in a calendar application or reminder service that can advise when and when not to send messages
- **Application usage** When and for what purpose specific applications are used
- **Preferences** User settings and reasons why they are chosen
- **Usage** Patterns that reveal how a device is used and the most effective time to send information
- **Network connectivity** When, how long, and why a user is connected to a network

This raw data can easily be distilled into relevant information that is helpful and timely to a user. Programs for accomplishing this task were developed about 10 years ago when the idea was first broached. However, the technology for pulling in all the needed data that would enable the programs to run with accuracy has yet to be conceived. The next generation of wireless networks could easily provide this information since the proposed always-on feature envisioned by proponents of the 4G networks accumulates an amazing amount of data by essentially keeping tabs on a user all the time.

Understanding the context of the wireless device user adds greater relevancy to the information that the user receives. Conversely, making sure that irrelevant information does not find its way into the subscriber's SMS inbox is an equally important service.

Of course, recognizing the downside of contextual information is easy, given the description of the information-gathering capabilities the devices possess: a lack of privacy.

Privacy in Wireless Messaging

Every new industry involving high technology raises concerns about consumers' privacy and the ability of organizations or businesses to invade that privacy. The use of wireless communications devices makes such invasions of privacy a real possibility. Some additional dangers include hackers gaining access, the misuse of personal information, unwarranted location tracking, and untargeted, unwanted messages with little or no relevance to the wireless subscriber's needs or interests (such as spam).

The mobile phone seems to be the ideal avenue for reaching consumers with personalized and targeted commercial messages. The concern is that the technological applications that make this targeted marketing possible could, if misused, result in a serious loss of privacy rights for the consumer.

As expected, the largest privacy concerns relate to location and contextual information gathering. Similar consumer fears about other information-gathering techniques have made headlines over the years. The ability to collect information about a user's travels on the Internet has been a major focus of privacy advocates. Other methods such as grocery store discount cards that enable the store to track every item purchased by customers have also come to light. The wireless networks' ability to gather such large amounts of personal information worries many consumers.

Established advertisers and marketers realize that consumers view this information gathering with a great deal of alarm. They also know that the products and services they market will suffer in the marketplace if consumers believe their privacy is being violated or their personal information is being misused. Consumers regard the wireless arena as even more warily than cyberspace because the devices used to amass information about them are personal and travel with them almost all of the time.

Self-Regulation of Wireless Advertising Failure to respond effectively to consumers' privacy concerns and the valid risks of information gathering could result in adverse consequences that range from consumer rejection of the technology to enforcement by a regulatory body or even costly litigation. The wireless industry associations recognized this potential threat some time ago. Currently, they are attempting to self-regulate by setting standards and practices that will alleviate consumers' fears while maintaining the marketers' ability to collect data needed to properly market their goods and services.

The *Mobile Marketing Association* (MMA), a worldwide organization committed to stimulating growth in wireless marketing and its associated technologies, has given membership to wireless agencies, wireless carriers and other service providers, advertisers, manufacturers, and retailers. In an attempt to establish privacy standards for the mobile marketing industry, MMA issued a list of guidelines designed to raise awareness of consumers' privacy concerns and encourage discussion among wireless carriers and marketers who practice location-based marketing.

The guidelines detail MMA's commitment to protecting the privacy of wireless subscribers while ensuring that mobile marketing campaigns remain effective and responsible. These guidelines, the first for privacy standards from an organization in North America, outline an environment for wireless users and wireless marketers that both parties feel is mutually beneficial. Some of the more important privacy solutions MMA recommends include the following:

- Members should not combine personally identifiable information with wireless subscribers' location information without the confirmed opt-in consent from the subscribers.
- Members that use anonymous or aggregate location information for marketing purposes should fully disclose the practice.
- Subscribers should be allowed to opt out of commercial programs at any time, even if they originally gave their consent to receive marketing campaigns.
- Members should have the consent of wireless subscribers before sharing subscriber information with third-party advertisers.

- Members that store subscriber information should implement stringent security measures to ensure that this data is not misplaced, misused, stolen, intercepted, or changed.

Another self-regulating group committed to protecting location privacy is the *Wireless Location Industry Association* (WLIA). This organization requires members to abide by privacy guidelines similar to MMA's guidelines, but it also requires each member to submit its privacy policy for certification. WLIA further ensures the protection of consumers' privacy by encouraging members to seek additional certification from other independent privacy policy review organizations.

One well-known independent privacy group is TRUSTe, a nonprofit certification organization that bestows its seal of approval on companies that agree to abide by its privacy guidelines. The TRUSTe seal is already a standard on the Internet, and the organization has begun setting similar protection standards for the wireless communications industry.

To this end, MMA and TRUSTe have entered into an agreement through which MMA is granted a seat on the TRUSTe Wireless Privacy Committee. TRUSTe can work with MMA member companies and include them in subsequent privacy initiatives.

It is unknown how TRUSTe intends to expand its seal-of-approval program to wireless communications marketing because the messaging medium involved is limiting. For example, an SMS message consists of only 160 characters, leaving little or no space for a privacy seal certifying that a TRUSTe member is sending the message. The EMS and MMS message formats contain more room so inclusion of such a seal becomes less of an issue as wireless networks advance their technology.

With any certification, enforcement constitutes a critical issue. Indeed, there is little point in establishing certifiable standards if there is no way to ensure a company's continued compliance with them. In the past, TRUSTe has been criticized for lacking the manpower to make its Internet guidelines stick. TRUSTe and similar organizations must overcome this perception if consumers' privacy fears are to be laid to rest.

Privacy Legislation Despite the proactive and self-regulatory actions that industry associations and independent organizations have taken regarding the wireless privacy issue, Congress has introduced several pieces of legislation that aim to protect consumer rights within the realm of wireless marketing:

- The Location Privacy Protection Act (July 2001) requires that wireless carriers and technology providers notify subscribers about their information collection policies regarding call-location data. The legislation requires that carriers also obtain subscriber consent before disclosing or selling such data.
- The Wireless Telephone Spam Protection Act (January 2001) prohibits the transmission of unsolicited marketing and advertisements to mobile phones and other wireless devices.
- The Wireless Privacy Protection Act (January 2001) requires full and clear notice by wireless carriers of their disclosure practices regarding customer location and transaction information. It also requires written subscriber consent to collect and use that information.

Wireless carriers and users must be equally aware of the threat to privacy that is inherent in a wireless device. It can report your location, your calling habits, whom you call, and your most frequent times for placing calls. It can also record where and how often you travel. When the next-generation networks arrive, wireless devices will be able to collect even more information. This powerful information deserves to be guarded diligently by consumers. They should not underestimate the need for conscientious regulatory bodies committed to their protection.

Voice and Wireless Messaging

Throughout this chapter, the discussion has been centered on using a wireless phone to send text messages, sounds, video, images, and files and surf the Internet. But what about voice calling, the most common application for wireless phones? It would seem that technological advances in telephony have resulted in the obsolescence of voice.

Not so. Voice continues to prosper. An overwhelming share of activity on wireless communications devices takes the form of voice calls. All the same, the substantial amount of information with which we find ourselves constantly bombarded and the ability to send and receive so many different types of data any time and any place seem to have minimized the role of voice telecommunications.

E-mail, faxes, and text message services demand our attention everyday. Because the newer technologies seem more interesting and are more difficult to master, these media tend to consume more of our time and energy. Voice still plays a major role; however, now it constitutes just one part of a greater communications whole. So, how do we manage that communications whole and what part does voice play in that management?

Unified Messaging (UM)

The variety of wireless communications methods currently in use can make managing the sheer volume of messages and information we receive difficult, inconvenient, and tedious. The time and effort involved in moving from point to point to retrieve messages can reduce productivity and cause overwhelming stress. In order to routinely pick up e-mail, for example, we must dial in, log on, or do both. When we are away from our terminal, we have to find another computer to collect our messages, if remote access is even possible. Sometimes we must wait for a fax, pick up faxes at the fax machine, or find a different fax machine to transmit a document if the recipient happens to be away from the office. When we want to check our voicemail, we have to dial in, accessing messages individually to determine the relevancy of the information, which is at times a slow process.

Now a service exists that consolidates all of this information by linking it to a single location—UM. UM is an emerging value-added network service that will soon generate more frequent text messages in order to become mainstream. When a new message arrives in the UM inbox, users receive a text message that includes an indication of the type of message (such as voicemail, e-mail, or fax) that is waiting. Messages can also be categorized by sender, priority, and time received. UM is much more convenient for users because it provides a single interface where they can access many different kinds of messages. This ensures that messages take precedence over the method by which they were sent.

As a component of a UM system, contextual information can help to choose the most appropriate message delivery method. For instance, when a message is sent to a subscriber and the application can ascertain that the recipient is away from his or her desk, the message can be forwarded immediately to the subscriber's mobile phone, through 'find-me, follow-me' technology. However, if the application determines that the recipient is at his or her desk, the message can be forwarded to the recipient's e-mail inbox. This leads to a more efficient method of making sure that information gets to the intended recipients while it is still relevant.

Most UM services centralize the user's mailbox on a web page accessed by the subscriber via a computer or WAP device. When a subscriber receives a text message on his or her wireless device announcing another message in the UM inbox, he or she checks the centralized mailbox. It might contain several messages. These are usually listed in the familiar e-mail message format with a header noting the time, date, and, of course, the medium through which the message arrived. Oddly, most UM services fail to include SMS messages in the UM inbox and only use SMS as a notification tool.

The means of accessing the message can take several different forms, depending on the original message format. The ability to perform efficient media and data conversion plays a major part in this process. Media conversion is often based on user preference and the level of the technology the subscriber is using. For instance, a fax might be converted into standard text e-mail and be accessed in that manner, or it might actually be converted into an image (such as a .gif or .tif file) and displayed using image-viewing software or as an HTML image.

Conversion may also be used on voicemail messages within the UM inbox. Most current computers can stream or store and play audio files so a voicemail message can be delivered and accessed in essentially its original form. If the device that the subscriber is using is incapable of playing audio, a speech-to-text conversion program can convert the message to a text message that resembles a standard e-mail message.

Users can sort, file, and store messages after accessing them and delete ones they no longer want in the same way as they do in most web-based e-mail applications. This feature eliminates the need to remember where a particular message is stored (on the mobile phone, in the e-mail program, in the voicemail box, in a fax somewhere in the office, and so on) because all the messages are routed through and stored in a central mailbox.

Despite the obvious advantages of UM services and the network providers' willingness to offer such a system, users have failed to leap at the opportunity to use this service. The reason might be simple: Many users maintain multiple accounts that they want to keep separate. Many people have a business e-mail address as well as a personal one, and unifying these accounts would defeat the purpose of having them in the first place.

In addition, media and data technology must overcome some obstacles to convert one type of message into another type. Much of this technology is either less than widely available, less than fully evolved (certainly to the level necessary for producing consistent results), or, as is often the case with new technology, unstandardized. Therefore, since essential software is limited, integrating all communications to a single access point will remain more conceptual than actual for the immediate future.

Voice Recognition (VR) Technology

In addition to the uses of voice within UM services (such as voicemail storage and speech-to-text software), other voice applications will become commonplace as voice and wireless technologies mature. As the number and variety of mobile services continue to expand, the need for an easier-to-use, more intuitive, and more effective user interface for wireless devices has become paramount. Subscribers are less likely to adopt increasingly complex messaging services if the user interface is so complicated that the time spent learning how to operate the device is perceived as not worth the benefits gained. In fact, many end users have already reached that point.

The wireless technology industry is addressing this growing consumer demand by pushing forward with technology that taps the most basic of user-input media: the human voice. *Voice recognition* (VR) technology will make interacting with wireless applications much easier and provide special dialing features, e-mail and phone list accessibility, message dictation, and voice web browsing and menu scrolling.

With this goal in mind, One Voice Technologies has developed what it claims is the industry's first voice-to-text messaging platform that enables mobile phone users to send e-mail, SMS, and pager messages using only their voice to enter the information. Other interface systems have been developed to understand words, concepts, and phrases by storing previous commands the user has made and, if necessary, asking the user clarifying questions. Many of these systems are currently undergoing testing or have actually been deployed in some of the larger wireless markets.

Commercial Applications of VR Technology Although users whose wireless messaging activity consists mostly of P2P exchanges will undoubtedly welcome VR technology, the business community will probably reap the most benefits. Users can enter product selections, payment information, shipping preferences, and more into the system through voice commands, creating a much less complicated interaction between customer and vendor.

VR technology will also bring additional benefits to businesses in the area of security. Phone access for commercial transactions or other critical information will be better secured through VR by using a voiceprint as a biometric measure. Because the identifying features go beyond a picture ID, Social Security Number, password, or *personal identification number* (PIN) to something biological (which is impossible to reproduce), this method renders those organizations and individuals who would misuse personal information powerless. When it is properly implemented, it will make phone transactions safer than most other types of business dealings.

The financial sector seems most able to gain from this added security in the short run. For example, users who want to access their bank account or stock portfolio via a wireless device will be able to enroll their voice with a PIN or other information such as an account number or password. When they attempt to access their accounts, the VR technology will compare the PIN as well as the voice entering the

PIN to the biometric database. This feature adds a second layer of security through more advanced identification techniques that ensure what the users know (their PIN, password, or account number) and who they are (their voiceprint).

The combination of user information and biometrics provides exceptional security that will be difficult, if not impossible, to circumvent. Furthermore, the use of voice in tandem with other forms of user input will drive the progress toward simplification of the user interface for wireless technology services. This fact seems more apparent than ever in the field of multimodal wireless services.

Multimodal Wireless Increasingly, manufacturers and communication industry leaders have focused on multimodal services in an effort to provide a next-generation wireless interface that is more user friendly. Multimodal services are communications services that users can access and use through multiple methods that include traditional input such as pressing keypad buttons with the additional option of speaking voice commands. This process enables service users to employ the most appropriate method at any particular time.

Multimodal applications combine these voice and touch input methods with text, audio, and graphic output, thereby combining the input and output modes in the user interface of a device. The resulting interface—touch screen, keypad, and voice commands for data input or sounds and images for output—allows easier interaction between the device user and device.

The multimodal interface in wireless phones, PDAs, and car navigation systems enables users to request information by voice, receive visual responses, interact with a stylus or the traditional keypad, or combine any of the input and output methods they choose. The option to select the best method of interaction for their needs and the ability to change methods at any time during the transaction will fuel adoption and usage rates of the new wireless services.

Multimodal Standards and Opportunities For an emerging technology to succeed in the wireless world, standards must be developed and applied so applications can work on all devices connected to all networks. To achieve this outcome, six wireless and computer industry leaders (Intel, Microsoft, Comverse, SpeechWorks, Cisco

Systems, and Philips Speech Processing) have collaborated to develop standards that will aid in the widespread adoption of multimodal services and applications.

The resulting organization, the *Speech Application Language Tags* (SALT) Forum, is committed to promoting the adoption of multimodal interaction with services and applications based on HTML, *eXtended Hypertext Markup Language* (X-HTML), and *eXtended Markup Language* (XML). Because the SALT Forum is an open industry initiative, industry leaders who comprise the organization will support and share the specifications and standards developed. This effort should lead to significant market penetration and adoption when the technology is mature and available.

The current multimodal model features a platform- and device-independent standard that will provide access to information, applications, and web services from home and office computers, telephones, PDAs, and a host of other wireless devices. The broad range of applications and wide choice of devices by which to access these applications should encourage consumer adoption of the technology while providing new revenue opportunities for the service providers. The providers are so positive of this that a recent industry report estimated that 282 million people will be using multimodal portals and services by the end of 2006.

Again, wireless industry experts cite wireless messaging services (SMS and, eventually, EMS and MMS) as the key to increasing revenue and consumer adoption of voice and multimodal technologies.

Wireless Voice Portals Multimodal and voice technologies have enabled service providers to offer wireless voice portals. These web-like portals provide rapid access to all types of information in a similar way as a web portal; however, the services are generally activated and used by voice. Simple voice commands navigate the user through the voice portal like a mouse or keypad navigates a user through a web portal.

As voice portal technology advances, users will be able to request information through the use of natural language ("Please locate movie listings and show times for the New York metro area.") as opposed to a system of keyword activators ("movies, listings, show times, New York"). Users will navigate a menu through voice commands, narrowing down the information requested more efficiently. Some portal providers are even experimenting with voice response systems with built-in "personalities" that reply to the user questions in natural language. Such systems are designed to reduce the user's awareness that he or she is interacting with a machine while making such interaction less stressful.

VR Handsets The predominant VR and multimodal applications are the ones users employ once they are connected to the wireless network. However, VR technology also has the potential to make the connection process easier through a phone without a keypad, which is a whole new breed of mobile phones.

With Telespree's VR handset, a user simply turns on the mobile phone and speaks a number into the device. The VR application then dials and connects the number. A speed dialer can also process the call when the user requests a name recorded in the built-in address book. Of course, the usefulness of such a device will be greater once multimodal interfaces become more common; however, in the short term, other applications based on this model could be extremely practical on current devices. Automotive applications (such as voice-activated onboard navigation and voice-activated mobile phone connection) have already caught on with luxury car manufacturers, who report that users value these devices not only for their convenience and ease of use, but also for their safety implications.

VR Technology and SMS

One of the greatest impacts VR technology will have on the wireless market will occur in SMS and its successors. The new technologies embody the potential to affect SMS on both ends of the message by giving the sender of the message several options when creating it and offering the recipient several choices when receiving it.

Senders of SMS text messages can continue to use the traditional keypad, but will enjoy the added ability to speak their message. The service provider's voice-to-text applications will convert the message to the standard SMS format and display it on the recipient's wireless device. The latter process is easier and takes much less time than the former.

VR technology features for SMS will bring new ease-of-use preferences and also open the application to individuals without a mobile phone. Senders without a mobile phone can transmit SMS messages in much the same way because wireless carriers will maintain a number that users can dial from any landline telephone. They would simply enter the wireless number of the message recipient and speak their SMS message. Again, the message will be converted from voice to text and processed to the recipient's wireless device.

Recipients of SMS messages will also have options. The traditional method of receiving SMS messages will still exist. In addition, subscribers away from their mobile phone or outside their coverage area will be able to call a number supplied by their carrier, who will convert their SMS text messages to voice with text-to-speech applications.

In addition to being able to send SMS messages, individuals without wireless devices will also be able to receive them. If someone without SMS service receives an SMS text message, the recipient's phone will ring, and he or she will enter an identification code. The carrier will then convert the text message to a voice message and deliver it.

Landline SMS SMS appeal has risen to such heights that it has created a demand for access to the services from landline telephones. Some of the services are already in place, such as a non-SMS subscriber's option to send a message to a subscriber's mobile phone via an intermediary service. Other proposed landline SMS include SMS-capable landline telephones and terminals that would enable landline users to send and receive SMS messages directly, bypassing the intermediary service. These phones and terminals would also enable landline SMS users to send messages to other landline SMS users, fax machines, and e-mail accounts.

Germany's Deutsche Telekom has already introduced a landline SMS pilot trial. The first phase enabled subscribers to the company's mobile network to send text messages to landline subscribers. The second phase was set during the summer of 2002. This phase enables users to send SMS messages from landline devices to Deutsche Telekom mobile subscribers and other SMS-enabled landline subscribers. In the future, messages from the landline devices will be transmitted to competing companies' wireless networks.

Another Deutsche Telekom system feature will enable customers without an SMS-enabled handset to have all messages they receive converted to voice and deposited in their Deutsche Telekom voicemailbox as a standard voicemail message. To bring all of these applications together, the company has already introduced a new cordless landline telephone that is SMS-enabled, partially returning the short message format to its mobile roots.

Other wireless carriers and hardware manufacturers are introducing landline SMS services and devices throughout Europe and Asia. The amazing popularity SMS has enjoyed in these markets since its introduction certainly bodes well for the adoption of the landline services. As these technologies improve and mature to encompass EMS and MMS, landline telephones might evolve in much the same way mobile phones until they are also referred to as *devices* or *communication appliances*.

Because North America lags behind these two markets in SMS usage overall, little or no pent-up demand exists for these services on landline telephones in the home or office. As a result, consumer demand might take several years to reach a level where U.S. carriers will offer them. The first phase of landline SMS would probably take the form of the intermediary model that uses an *interactive voice response* (IVR) system to create and send the SMS message.

Interactive Voice Response (IVR) IVR systems have been in use for a number of years. Anyone who has ever called a large company, bank, movie theater, pharmacy, or even a local supermarket has encountered an IVR system. However, despite their familiarity, or perhaps because of it, consumers generally dislike them. Most consumers find them difficult or tedious to navigate. The result is often more frustration than satisfaction.

IVRs are cost effective for companies with automated tasks that can be performed without human intervention. The two types of IVR systems consist of *dual-tone multifrequency* (DTMF) signaling (the most common form), where callers respond to spoken questions by pressing buttons on a phone keypad, and voice response (the newer form), where a caller responds to spoken questions by voicing commands into the phone. The latter system is unreliable at times and has difficulty with different speech patterns and accents.

Since they are commonly used within voicemail services, IVR systems can easily be adapted for sending SMS messages to or receiving them from users without wireless service or without an SMS-enabled handset. IVR might become the mediating technology that provides access to many different types of wireless and voice-based services.

Wireless Number Portability (WNP)

The *Federal Communications Commission* (FCC) has mandated that *Wireless Number Porting* (WNP) and Pooling requirements commence on November 24, 2003, for the top 100 *metropolitan statistical area* (MSA) wireless carriers. Number portability refers to a circuit-switched telecommunications network feature that enables end users to retain their phone number when changing carrier, service type, location, or any combination of these elements. WNP will be phased in over a long period of time. After WNP has been fully implemented, a consumer might never have to change his or her phone number again. However, full implementation of the number portability protocols is years away. The three types of number portability include the following:

- **Service provider portability** Enables consumers to change local carriers while retaining the same phone number. This phase of number portability implementation is the only one currently being executed.
- **Location (geographic) portability** Enables consumers to change from one geographic location to another (for example, from New York to Texas) while retaining the same phone number. The current *Location Routing Number* (LRN) system permits some limited relocation within the rate boundaries. This phase of number portability implementation is still in the planning stage.

- **Service portability** Enables a subscriber to change services (for example, from CENTREX to POTS) while retaining the same phone number with the same carrier. This phase of number portability implementation is still in the planning stage.

Each of these phases of WNP implementation will affect subscribers and carriers in many and different ways.

What WNP Will Mean to the Wireless Carriers

Some WPN issues affecting carriers are technical in nature (such as mobility and roaming support), whereas others are industry-specific concerns related to regulatory matters (such as local rate centers versus market license areas and industry initiatives). Still other issues depend on successfully integrating the few differences between the wireline and wireless systems.

WNP will be *transparent* to consumers, meaning that calls to and from the ported subscriber will seem to connect in the same way they always have even though, on a technical level, major changes have taken place in the routing of these calls. The carriers, however, will bear the logistical burden of the greater intricacy in the routing, switching, and signaling process. Indeed, WNP can create a number of complex technical burdens and expenses for carriers because it requires most operators to complete numerous modifications to their systems in a relatively short time. Carriers are also involved in a series of mandated initiatives due for completion on or near the WNP deadline. These initiatives include *Enhanced 911* (E911), *Communications Assistance to Law Enforcement Act* (CALEA), and *text telephone* (TTY), not to mention new product or service rollouts the carriers might have planned.

Understandably, wireless service providers are apprehensive about the portability mandate. WNP implementation will increase the complexity of number administration, network operations, call processing, and service assurance, and will require changes in nearly every aspect of business operations from intercompany data exchange and call routing to customer service and billing. Any WNP solution will be required to integrate smoothly with the existing support systems to assist carriers in managing porting activities efficiently and reliably while continuing to provide dependable service to their subscribers.

What WNP Will Mean to You

Number portability will provide the greatest benefit to the wireless subscriber in the long run. The ability to take your phone number with you when you change geographic locations, change carriers, or change to a wireless service provides a giant gain in convenience. Remember, not long ago such an advantage was inconceivable. However, with the full implementation of WNP, it will be possible to keep a phone number an entire lifetime.

In the short run, however, the consumer might feel the brunt of the WNP implementation. Service-related problems such as E911 callback difficulties, delays in activation, the possibility that calls but not SMS messages could connect, or outright porting failures could conceivably result from such a massive and complex changeover.

Today, wireless messaging has truly reached global proportions. For the wireless world, the immediate future still resides in large part with SMS—but not for long. As wireless technology rapidly matures from SMS to EMS to MMS, the number and complexity of messaging applications continue to multiply. Indeed, we have stopped wondering whether wireless messaging can really last. With no end in sight for the foreseeable future, everyone's now wondering how far we can take it.

GLOSSARY

1G First-generation cellular. Analog cellular, including AMPS.

2G Second-generation cellular. Digital cellular, including TDMA, CDMA, and GSM systems. Most 2G digital phones are voice-only phones, but some offer limited data capability.

2.5G Enhanced data rate second generation. Digital cellular systems with data rates of 28 to 384 Kbps.

3G Third-generation wireless networks. The next generation of wireless network technology. These networks offer the promise of eventually being able to deliver voice, data, and multimedia content at rates as high as 2 Mbps.

Access point A stationary device that acts as a base station for WLAN users. Unlike an NIC that connects to a mobile device, the access point connects directly to a wired network.

Adaptive frequency hopping A method whereby a Bluetooth radio would first check that a band was clear before it attempted a transmission. This would enable Bluetooth radios to exist more peacefully with other radios such as 802.11b.

Adds Additions. The number of subscribers a carrier adds in a given period (monthly, quarterly, and/or annually). They are typically measured in terms of net adds (the number of additional subscribers minus the number that have churned) or gross adds (the total additions for that period).

AGC Automatic gain control.

Aggregation Aggregation is the process of collecting charges for multiple transactions and combining them on a single bill. Charges are typically aggregated when the cost of processing the individual transaction exceeds the profit that would be realized from that transaction.

AMPS Advanced Mobile Phone Service. Commonly known as analog cellular. AMPS service is available in the United States, Mexico, Canada, Australia, and several other countries. It is used in the 800 MHz frequency band.

ANSI American National Standards Institute.

Antenna diversity The use of two or more antennas to improve signal quality. In most designs, the baseband processor automatically selects the antenna that is providing the best quality signal.

ARPU Average revenue per unit or average revenue per user. (Both are common usage.) This refers to the amount of gross revenue a carrier can expect, on average, from its customers. Typically, this is computed on a monthly, quarterly, and annual basis.

ASP Average selling price.

ATM Asynchronous Transfer Mode.

Bandwidth A measure of the capacity of a communications channel and the amount of frequency available to a system. The wider the bandwidth allocated to a channel, the greater the data rate for a given protocol.

Base station A transmission and reception station for handling cellular traffic. It usually consists of one or more receive/transmit antennas, a microwave dish, and electronic circuitry. This is also referred to as a *cell site* since it holds one or more receive/transmit cells.

BBC Baseband Converter.

BFWA Broadband Fixed Wireless Access.

BiCMOS Bipolar complementary metal oxide semiconductor.

Bit rate The speed at which bits are transmitted over the physical layer. This is also called the signaling rate. This is quite different than throughput, which is an end measure of a network's speed.

Bluetooth A short-range transmission technology for multiple device networking.

BREW Binary Runtime Environment for Wireless. A technology developed by Qualcomm.

BS Base station.

BTS Base Transceiver Station.

CAGR Compounded annual growth rate.

Carrier The base unmodulated frequency used by a system. The modulation process will generate a signal centered on the carrier that has a width that is equal to the bandwidth.

CDMA Code Division Multiple Access. Also known as IS-95. This is one of the newer digital technologies in use in the United States, Canada, Australia, and some southeastern Asian countries (for example, Hong Kong and South Korea). CDMA differs from GSM and TDMA in its use of spread spectrum techniques for transmitting voice or data over the air. Rather than dividing the RF spectrum into separate user channels by frequency slices or time slots, spread spectrum technology separates users by assigning them digital codes within the same broad spectrum. Advantages of CDMA include higher user capacity and immunity from interference by other signals. It is used in either the 800 or 1,900 MHz frequency bands.

CDMA2000 2G+ CDMA including 1xRTT and 1xEV.

CDMA-One 2G CDMA, IS-95B.

CDPD Cellular Digital Packet Data. A protocol designed and deployed over analog WANs (typically AMPS cellular networks). CDPD transparently piggybacks on cellular analog conversations to enable simultaneous voice/data transmission.

Cell The basic geographic unit of a cellular system and the basis for the generic industry term *cellular*. A city is divided into small cells, which are equipped with a low-powered radio transmitter/receiver or base station. The cells can vary in size depending on the terrain and capacity demands.

Channel On the radio, this is usually a synonym of a specific frequency and, by extension, the communication medium. It can also mean a stream of data between two nodes (a point-to-point link in connection-oriented systems).

Chipset A group of IC chips that are designed to work together and are generally used and priced as a set.

CHTML Compact Hypertext Markup Language (the basis for i-mode).

Churn An industry term that refers to customer turnover. Wireless subscribers are said to churn when they cancel their mobile service with their current wireless carrier. Subscriber may either churn to another carrier, or they may simply choose not to have any wireless service. Churn is measured on a monthly basis. To get the total churn for a given period (typically quarterly or yearly), the monthly churn percentage is multiplied by the number of months in the period being measured. For instance, a carrier with 2 percent churn a month would have a quarterly churn rate of 6 percent and an annual churn rate of 24 percent.

Codec Compressor/decompressor. Refers to the hardware in a cell phone and in the cell network that compresses digitized voice prior to transmission and takes received compressed voice and decompresses it prior to passing it to either a cell phone speaker or into a wireline system.

Control channel A channel used for the transmission of digital control information from a base station to a cellular phone (forward control channel) or from a cellular phone to a base station (reverse control channel).

CSC Circuit-switched cellular. This is good for large data transfers and offers wide coverage.

CSMA Carrier Sense Multiple Access. A MAC method of listening before transmitting (collision avoidance) and listening while transmitting (collision detection). For a wired network, such as Ethernet, collision detection is employed and packets are retransmitted should a collision be detected while transmitting. For wireless networks, this type of collision detection is usually not possible since the strength of a radio's own transmissions would mask all other signals on the air. So for wireless networks, collision avoidance is employed.

D/A Digital-to-analog converter.

D-AMPS Digital AMPS.

dB (decibel) A logarithmic way to express a value. Usually, the signal strength (transmitted and received power) is expressed in dBm (the reference is 1 mW = 0 dBm).

DCS1800 European 1,800 MHz GSM band.

DECT Digital European Cordless Telephone.

DQPSK Digital quadrature phase-shift keying.

DSP Digital signal processor.

Dual band Dual-band phones are capable of using two different frequencies of the same technologies—for example, a TDMA or CDMA phone that can use either the 800 or 1,900 MHz band. There are also triple-band phones in the GSM market that support 900, 1,800, and 1,900 MHz. Dual-band phones enable callers to access different frequencies in the same or different geographic regions, essentially giving their phone a wider coverage area.

Dual mode Dual-mode phones are phones that support more than one technology. Typically, this is either CDMA and AMPS or TDMA and AMPS, but other dual-mode phones are starting to appear on the market, such as GSM and TDMA.

Duplexer A signal-separating filter. It incorporates two filters—one for the reception frequency and one for the transmission frequency—and separates those signals. There are also duplexers that include a switch.

E911 Enhanced 911. A service mandated by the FCC for U.S. mobile carriers. The service will enable location information for subscribers calling 911 to be transmitted automatically to a PSAP.

EDGE Enhanced Data Rates for Global Evolution. A high-data-rate (up to 384 Kbps), packet-based technology being developed for TDMA-based networks (such as IS-54, IS-136, and GSM).

EIRP Effective Isotropic Radiated Power. A measure of the radiated power from a transmitter entering the atmosphere.

E-mail Store-and-forward messaging that can contain attachments—typically 1,000 characters long.

EMS Enhanced Messaging Service. Messaging comprising simple pictures, sounds, animations, and modified text.

ERMES European Radio Messaging System.

ESMR Enhanced Specialized Mobile Radio.

ESN Electronic serial number. Each cellular phone is assigned a unique ESN, which is automatically transmitted to the cellular base station every time a call is placed. The MTSO validates the ESN with each call.

Ethernet Standard wired LAN protocol. It includes physical and link layers.

ETSI European Telecommunications Standards Institute.

Fading The variation in channel performance due to the dynamicity of the environment; this changes the receive signal strength.

FCC Federal Communications Commission. The governing body of radio in the United States.

FDMA Frequency Division Multiple Access.

FEC Forward error correction. A technique used to overcome some type of errors created by transmission on noisy channels by adding redundancy bits to the main data transmission.

Frequency band A portion of the radio spectrum delimited for a particular use. For example, most WLANs currently use the 2.4 to 2.48 GHz band, although 5 GHz band products are on the way. A frequency band is usually divided in two channels.

Frequency reuse The ability to use the same frequencies repeatedly across a cellular system. This is made possible by the basic design approach for cellular. Since each cell is designed to use radio frequencies only within its boundaries, the same frequencies can be reused in other cells not far away with little potential for interference. The reuse of frequencies enables a cellular system to handle a huge number of calls with a limited number of channels.

GGSN Gateway GPRS Service Node.

GHz Gigahertz (1,000,000,000 Hz).

GPRS General Packet Radio Service. A packet-switched data technology that is being deployed primarily for GSM networks.

GSM Global System for Mobile Communications. GSM is used all over Europe and in many countries in the Middle East, Asia, Africa, South America, Australia, and North America. GSM's air interface is based on narrowband TDMA technology, where available frequency bands are divided into time slots; each user has access to one time slot at regular intervals.

Handoff The process where the MTSO passes a cellular phone conversation from one RF in one cell to another RF in another. The handoff is performed so quickly that users usually never notice.

HCS Hierarchical cell structures.

HDML Handheld Device Markup Language.

Header Information added by the protocol in front of the payload in the packet for its own use (addresses, packet type, sequence number, CRC, and so on). Each protocol adds a different header, so in a typical TCP/IP packet, we have a MAC header, an IP header, and a TCP header, which is followed by the payload.

HSCSD High-Speed Circuit-Switched Data. This increases the GSM channel data rate up to 14.4 Kbps through improved compression technologies. By using multiple time slots, the channels can be multiplexed together to offer data rates up to 57.6 Kbps.

HTML Hypertext Markup Language.

IC Integrated circuit.

iDEN Integrated Digital Enhanced Network. A modified TDMA technology used by Motorola and run by Nextel Communications, Southern LINC, and a handful of other carriers around the world. iDEN phones run on a different frequency than other cellular services and are therefore incompatible with them.

IEEE Institute of Electronic and Electrical Engineers. The IEEE is a nonprofit, technical professional association that promotes electronic ideas and standards in the United States and worldwide.

IF Intermediate frequency. To increase sensitivity and selectivity, super heterodyne radio receivers first convert the input frequency to a fixed frequency and then apply the internal processing to that fixed frequency. The IF is the frequency used for this internal signal processing.

i-mode Wireless Internet service from NTT DoCoMo. It is based on CHTML.

IMT International Mobile Telephone (see UMTS).

IM Instant Messaging. Messaging with presence (the ability to view who is present). IM is capable in all network types.

IP Internet Protocol.

IPX The network protocol used in Novell Netware, usually with SPX.

IS-136 TDMA interim standard 136. (See TDMA.)

IS-95 CDMA interim standard 95. (See CDMA.)

ISM Industrial, Scientific, and Medical unlicensed frequency bands. Operates at 900 MHz and 2.4 GHz.

ITU International Telecommunication Union.

ITU-T International Telecommunication Union-Telecommunications Standardization Sector.

IVR Interactive voice response. A software application, typically used in conjunction with corporate telephony hardware, that recognizes spoken commands. Typically used for helping callers navigate corporate directories and phonebooks or for other types of menu-driven services. Usually limited in the number of commands that can be recognized.

J2ME Java 2 Microedition. A technology developed by Sun Microsystems.

KHz Kilohertz (1,000 Hz).

LAN Local area network. A network covering short distances, such as within a building.

Layer This usually refers to the OSI specification dividing any communicating system into seven layers, which have different functionalities. Layer 1 is the physical layer (radio), layer 2 is the link layer, and IP could be assimilated as layer 3 (network layer). TCP is considered layer 4, the transport layer.

LBS Location-based services. These are services or applications that center on a user's location in a mobile environment. LBS utilize location-sensitive technology, such as GPS or network-based solutions, to deliver services or applications to a wireless device such as a mobile phone. These services can include finder applications that let mobile phone users locate friends or family, businesses, or landmarks. They can also deliver maps, directions, or traffic reports.

Link layer This is the part of the protocol managing the direct delivery between two devices on a specific physical layer (coaxial bus, point-to-point link, and radio). This includes packetization and addressing. Most of this is implemented in the MAC in a WLAN.

LMDS Local Multipoint Distribution Services. LMDS broadband services operate over the 28 to 31 GHz bands in the United States and provide high data rates, but only over a relatively short distance of three miles.

LNA Low-noise amplifier.

Location-based commerce Location commerce (l-commerce) refers to commercial transactions that take place in a mobile environment, but are dependent in some way on the physical location of the customer or the physical location of a business.

MAC Medium Access Control. This is the part of the radio device managing the protocol and the usage of the link. The MAC decides when to transmit and when to receive, creates the packets headers, and filters the received packet.

MBS Mobile Broadband System.

M-commerce Mobile commerce or mobile electronic commerce. This refers to commercial transactions and payments conducted in an untethered, non-PC-based environment. Transactions are made using wireless devices that can access data networks and send and receive information, including personal financial information.

MCD Mobile computing device or multicommunication device.

Micropayments Small payments that are typically aggregated by an m-wallet provider or other payment processor. Cahners In-Stat/MDR considers payments from $.01 to $2.00 to be micropayments.

Micropayment aggregation Micropayments are aggregated because the cost to process each micropayment individually may exceed the total cost of the transaction. Typically, processing charges on amounts below $9.99 is not cost effective. Therefore, In-Stat believes that carriers and merchants will aggregate all micropayments and charges below $9.99 or turn to third-party aggregators.

MMDS Multichannel Multipoint Distribution Service. A fixed wireless service for data, voice, and video that operates in the 2.5 GHz band in North America and in the 3.5 GHz bandwidth internationally.

MMS Multimedia Messaging Service. A type of messaging comprising a combination of text, sounds, images, and video.

MNO Mobile network operator. An operator of a wireless network for mobile phones.

Modem Modulator/demodulator. A radio device. This is the part that converts the bits to transmit into a modulation of the radio waves and the reverse at the reception. It performs the analog-to-digital conversion, the generation of the frequency, the modulation, and the amplification.

mW Milliwatt (.001 watt).

MTSO Mobile Telephone Switching Office. The central switch that controls the entire operation of a cellular system.

M-wallets Mobile wallets. These are software applications that hold a user's sensitive personal and financial information, such as credit card numbers, expiration dates, bank account information, passwords, and PINs. Most m-wallets are server based, which theoretically is more secure and avoids placing data onto mobile devices, which are often processor and memory constrained.

MVNO Mobile virtual network operator. A company that to end users appears to be a wireless network operator. Unlike a standard wireless carrier, however, an MVNO does not own the BSS that an MNO owns. MVNOs also may not necessarily own other infrastructure that is normally associated with an MNO, such as MSCs and HLRs. More importantly, MVNOs do not hold licenses to radio spectrum; instead, they purchase network capacity from wireless carriers that hold licenses and operate the network infrastructure necessary for wireless phone communication.

NAM Number Assignment Module. The electronic memory in the cellular phone that stores the telephone number. Phones with dual- or multi-NAM features offer users the option of registering the phone with a local number in more than one market.

NIC Network interface card. Otherwise known as a WLAN card. In most cases, this board or PCMCIA device is added to a computer or portable device to give it WLAN capabilities, but increasingly, manufacturers are incorporating network interface circuitry into portable devices, thereby eliminating the need for a separate NIC.

NMT Nordic Mobile Telephone.

Noise Any unwanted signal. This can include background noise, interference, or transmissions from nodes that do not belong to the network. See also SNR.

Number portability The ability for wireless subscribers to retain their mobile phone number when they switch mobile carriers.

PA Power amplifier.

Packet A unit of transmission over a network. The data to be transmitted is split into packets, which are sent individually over the network.

Paging The act of seeking a cellular phone when an incoming call is trying to reach the phone.

PBCC Packet binary convolutional coding. An optional modulation scheme that is part of the 802.11g standard.

PCMCIA Personal Computer Memory Card International Association. A PC interface card standard used in mobile devices.

PCN Personal Communication Network. This standard corresponds to a high-frequency version of the GSM standard.

PCS Personal communication services. PCS and cellular are sometimes interchanged. Officially, PCS is a digital cellular service in the 1,900 MHz band only. In practice, some providers have used part of their AMPS 800 MHz allocation to offer PCS or digital cellular services.

PDA Personal digital assistant. A handheld computer that can be used for simple PIM functions. As PDAs have become increasingly sophisticated, some are poised to usurp the place of subcompact notebook computers and can be used for more complex functions such as viewing and editing documents, spreadsheets, and presentations. PDAs run on a variety of operating systems, the most common of which are the Palm, Microsoft PocketPC, Symbian, and Linux.

PDC Personal Digital Cellular (Japan).

PHS Personal Handy Phone System (Japan).

Physical layer The part of the device interacting with the medium. For a WLAN, the physical layer is the radio.

PIM Personal Information Manager.

PIN Personal Identification Number.

PLL Phase locked loop. Circuit technology (or a circuit using that technology) in which the circuit is operated at an arbitrary frequency by forming a loop circuit that synchronizes the frequency phase.

PM Phase modulation.

POS Point of sale terminal. A device that accepts credit/debit card payments.

Presence Classifying one as available for correspondence (must be available).

Protocol A specification of the interactions between systems and the data manipulated. This describes what to do and when (the rules) to do it, and the format of the data exchanged on the lower communication layer.

PSAP Public Safety Answering Point. The state-funded call center that receives all 911 calls and routes the calls to the appropriate emergency agency.

QoS Quality of service. A measure of how reliable a carrier's service is. This is usually expressed in terms of availability and is measured, as how often available, by .99999 (or five nines), which is the top level of reliability.

RACE Research and Development in Advanced Communications Technologies in Europe.

RAM Random Access Memory.

Registration The procedure that a cellular phone initiates to a base station to indicate that it is active.

RF Radio frequency.

RLL Radio in the local loop.

Roaming The ability to move between cells of the same network. Also the ability to use a cellular phone outside a provider's home service area. Providers often set up roaming agreements with other providers in different geographic locations. A roaming agreement lets a caller seamlessly make calls in the other provider's geographic service area without operator intervention.

ROM Read-only memory.

SAW filter Surface Acoustic Wave filter. A filter that uses surface elastic waves that are transmitted across the surface of a piezoelectric material. This implements a filter that has the resonant frequency and its vicinity as the pass band.

SGSN Server GPRS Support Node.

SID System identification. A five-digit number that indicates which service area the phone is in. Most carriers have one SID assigned to their service area.

SIM Subscriber Identity Module.

SIM card A small memory card not much bigger than half the length of a person's thumb. This is used in GSM phones to hold phone numbers and other information. It can be removed and inserted into other GSM phones, enabling callers to keep their numbers and place and receive phone calls.

SIM toolkit A standard for value-added wireless services that enables the end user to establish an interactive exchange with network applications.

SMS Short Message Service. A method of delivering a short (120 to 160 character) message to a digital cellular phone. GSM phones can also send SMS messages. It is network provider dependent.

SNR Signal-to-noise ratio. The difference in strength between the signal we want to receive and the background noise (or any unwanted signal).

SP-lock A lock placed on a cellular phone by some service providers to ensure that subscribers can only use the phone with the carrier's service.

SSL Secure Socket Layer.

Standby time The amount of time callers can leave their fully charged cellular phone turned on before the phone will completely discharge the batteries.

Speech recognition A software application that can recognize spoken speech—a human voice. Voice recognition applications do not understand what the content of the speech means; however, they can recognize specific commands, such as "Call Jane Smith," and perform specific actions based on those commands. Over the last few years, speech recognition technology has improved dramatically. Early efforts were marked by poor accuracy and a need to speak slowly. Current products recognize many different commands, speaking styles, and accents.

TACS Total Access Communications System.

Talk time The length of time callers can talk on their cellular phone without recharging the battery. The battery capacity of a cellular phone is usually expressed in terms of so many minutes of talk time or so many hours of standby time.

TCP Transmission Control Protocol.

TCP/IP Network protocol used by Unix and Internet. This is better in some respects than NetBeui and IPX (it allows routing, for example).

TDMA Time Division Multiple Access. A technique used to share the same bandwidth between different channels using periodic time slots. TDMA divides frequency bands available to the network into time slots; each user has access to one time slot at regular intervals. TDMA thereby makes more efficient use of available bandwidth than the previous generation AMPS technology. It can be used on either 800 or 1,900 MHz frequency bands.

Telemetry See wireless telemetry.

Throughput A measure of the performance of a network for large data transfer (such as FTP, NFS, and HTTP 1.1). This speed is expressed in bits per seconds or a multiple.

TM-UWB Time-Modulated Ultra Wideband. A method of wirelessly sending data that indicates 1s or 0s by varying the time between ultrafast pulses.

Transceiver See wireless transceiver.

Tri-band A phone that can operate on three bands. This is typically a GSM phone operating on 900, 1,800, and 1,900 MHz.

TTS Text to speech. The flip side of speech recognition. TTS takes written words and converts them to speech. Thus, when a caller requests specific information from a voice portal, such as driving directions, TTS reads the directions to the caller. Early TTS efforts were slow and were usually read by a computerized voice that was often referred to as *Igor* because of its similarity to the voice of the character of the same name in old horror movies. Current TTS technology is much more natural sounding, and in some situations it would be difficult for the caller to differentiate TTS from an actual human speaker.

UMTS Universal Mobile Telephone System.

U-NII Unlicensed National Information Infrastructure band. Operates from 5,725 MHz to 5,825 MHz.

UWB Ultra Wideband. A method of transmitting information that encompasses a large portion of the radio spectrum.

UWC Universal Wireless Consortium.

VCO Voltage-controlled oscillator. An oscillator in which the output frequency varies according to an input control voltage.

Voice channel The channel the switch assigns to a caller to commence the call on after the exchange of subscriber data.

Voice portal A voice portal is a software application that uses speech recognition technology to provide information to callers. Using a combination of speech recognition and TTS technology, the application lets callers request specific information, such as news, weather, traffic reports, or e-mail, which is read by the application, to the caller. Voice portals can also enable callers to conduct transactions, such as trade stocks or manage bank accounts. Callers can also use voice portals to purchase products or services. Voice portals essentially enable callers to perform functions that they might otherwise do using the Internet or other methods. Additionally, the application can be used to authenticate callers by matching their voiceprint to one on file for security purposes.

VoIP Voice over IP.

VXML Voice eXtensible Markup Language. The standard Internet markup language for use in speech applications. It enables voice portal applications to access Internet content and read it to callers.

WAP Wireless Application Protocol. A standard or protocol for wireless devices and the accompanying infrastructure equipment.

WAP provides a standard way of linking the Internet to mobile phones, PDAs, and pagers/messaging units.

WAP-NG WAP Next Generation.

W-CDMA Wideband CDMA. A 3G evolutionary path for GSM and TDMA technology.

WEP Wired Equivalent Privacy. An algorithm whereby a pseudo-random number generator is initialized by a shared secret key. When this encryption is incorporated into a WLAN, eavesdropping is much more difficult.

Wireless telemetry Wireless machine-to-machine (M2M) communication, not including traditional data- or voice-centric devices. Examples of wireless telemetry applications include asset tracking, point of sale, vending, arcade games, supervisory control and data acquisition (SCADA) such as energy/utility, and traffic-monitoring applications.

Wireless transceiver A wireless device (typically a modem) that transmits M2M data from the unit being monitored to a control room where it can become useful information.

WPOS Wireless point of sale. Wireless M2M communication, not including traditional data- or voice-centric devices, that allows for credit/debit card transactions. Typically, receipt capabilities exist. This does not include credit/debit card purchasing capabilities through a traditional voice-centric handset.

WML Wireless Markup Language.

WTLS Wireless Transport Layer Security.

X-HTML eXtended Hypertext Markup Language.

XML eXtended Markup Language.

INDEX

Symbols

A

B

C

D

E

F

G

H

I

J–K

L

M

N

O

P

Q

R

S

T

U

V

W

X

ABOUT THE AUTHORS

Donald Longueuil (author) is a Wireless Analyst for the research and consulting firm, In-Stat, a subsidiary of Reed Elsevier Publications. For almost 4 years, Mr. Longueuil has been an advisor to CEOs, senior management, and investment professionals on global telecommunications issues, identifying and analyzing trends related to telecom operators. Mr. Longueuil began his career with The Yankee Group. He has been widely published in leading industry and business periodicals, as well as contributing articles to numerous trade journals.

Sander Brouwer is a telecommunications specialist with over eight years of experience in wireless messaging. Mr. Brouwer is the Global Product Manager of CMG's industry leading Short Message Service Center, and holds a degree in IT and Telecommunications. He is a frequent speaker on congresses and has written multiple articles on messaging and related areas.

Colin Matthews (co-author), President and CEO of Infomatch, brings more than 20 years of experience in establishing, growing, and managing high-tech companies. Previously a Vice President and Corporate Officer with Sedona Corporation (NASDAQ: SDNA) and President of the firm's software division, Sedona GeoServices, Colin assumed responsibility for turning the company around and positioning it as an Application Service Provider (ASP), serving the Business Intelligence/ERP and information delivery markets.

Geof Wheelwright (co-author) is a journalist and author with 22 years of experience in covering the telecommunications and high technology sector. He is a regular contributor to the Financial Times, as well as NASDAQ International Magazine and Canada's National Post. In addition, he regularly writes for a number of British and North American trade journals, and provides consulting services.